Recent Titles in This Series

(Continued in the back of this publication)

MEMOIRS

of the
American Mathematical Society

Number 533

Anticipative Girsanov Transformations and Skorohod Stochastic Differential Equations

Rainer Buckdahn

September 1994 • Volume 111 • Number 533 (second of 5 numbers) • ISSN 0065-9266

American Mathematical Society
Providence, Rhode Island

1991 *Mathematics Subject Classification.*
Primary 60H07; Secondary 60H10.

Library of Congress Cataloging-in-Publication Data

Buckdahn, Rainer.
  Anticipative Girsanov transformations and Skorohod stochastic differential equations / Rainer Buckdahn.
      p.   cm. – (Memoirs of the American Mathematical Society, ISSN 0065-9266; no. 533)
  Includes bibliographical references.
  ISBN 0-8218-2596-8
  1. Stochastic differential equations. I. Title. II. Series.
QA3.A57   no. 533
[QA274.23]
510 s–dc20   94-17087
[519.2]   CIP

Memoirs of the American Mathematical Society

This journal is devoted entirely to research in pure and applied mathematics.

Subscription information. The 1994 subscription begins with Number 512 and consists of six mailings, each containing one or more numbers. Subscription prices for 1994 are $353 list, $282 institutional member. A late charge of 10% of the subscription price will be imposed on orders received from nonmembers after January 1 of the subscription year. Subscribers outside the United States and India must pay a postage surcharge of $25; subscribers in India must pay a postage surcharge of $43. Expedited delivery to destinations in North America $30; elsewhere $92. Each number may be ordered separately; *please specify number* when ordering an individual number. For prices and titles of recently released numbers, see the New Publications sections of the *Notices of the American Mathematical Society.*

Back number information. For back issues see the *AMS Catalog of Publications.*

Subscriptions and orders should be addressed to the American Mathematical Society, P. O. Box 5904, Boston, MA 02206-5904. *All orders must be accompanied by payment.* Other correspondence should be addressed to Box 6248, Providence, RI 02940-6248.

Memoirs of the American Mathematical Society is published bimonthly (each volume consisting usually of more than one number) by the American Mathematical Society at 201 Charles Street, Providence, RI 02904-2213. Second-class postage paid at Providence, Rhode Island. Postmaster: Send address changes to Memoirs, American Mathematical Society, P. O. Box 6248, Providence, RI 02940-6248.

10 9 8 7 6 5 4 3 2 1   99 98 97 96 95 94

Contents

Abstract

The theory of the stochastic integration of processes that are not necessarily adapted to the Wiener process has been recently developed by several authors. In particular, D. Nualart and E. Pardoux [33] have developed a practicable extended stochastic calculus for both the Skorohod integral - an extension of the Itô integral - and the Stratonovich integral - an extension of the Itô-Stratonovich integral. This theory allows to study stochastic differential equations for which the solution is a nonadapted stochastic process. The theory of stochastic differential equations with Stratonovich integral is well-studied, cf., e.g. D. Ocone/E. Pardoux [38] and D. Nualart/E. Pardoux [34]. The main method used here consists in substituting finite variation processes depending on the whole path of the Wiener process in the adapted flow associated to the stochastic differential equation.

This method does not work for stochastic differential equations with Skorohod integral. So earlier attempts to solve these equations made use of the Wiener chaos decomposition, cf. Y. Shiota [42] and A.S. Ustunel [44]. However, this method is restricted to linear equations with deterministic coefficients and possibly random initial value. On the other hand, stochastic differential equations with Skorohod integral can be understood as hyperbolic first order partial stochastic differential equations with Stratonovich integral, where, in addition to the differential of the Wiener process, the Fréchet derivative of the solution with respect to the Wiener process occurs, too. The method of solving such stochastic differential equations consists in eliminating the Fréchet derivative by a Girsanov transformation. Of course, this Girsanov transformation is also needed for shift processes not adapted to the Wiener Process.

Mainly basing on the author's previous papers [4], [8], [10], [11] and [12] and on common works with D. Nualart [16] and H. Föllmer [15], the present Seminarbericht studies anticipative Girsanov transformations over the standard Wiener space and stochastic differential equations with Skorohod integral. The study of the anticipative Girsanov transformations is not only restricted to the presentation of a tool for tackling stochastic differential equations with Skorohod integral: Motivated by the papers of R. Ramer [41], S. Kusuoka [29] and D. Nualart/M. Zakai [36], which follow earlier work of R.H. Cameron and W.T. Martin, L. Gross and H.H. Kuo, different types of anticipative transformations are considered and sufficient conditions for their absolute continuity with respect to the Wiener measure are derived.

Although the paper bases mainly on the papers quoted above, it is not only a mere summary of them: Some new ideas are presented and relations between the papers are utilized for abbreviating some derivations.

The Seminarbericht consists of two main chapters, Chapter 2 on anticipative Girsanov transformations and Chapter 3 on stochastic differential equations with Skorohod integral. Taking into account the earlier separate development of both subjects, each of the two chapters is complemented with an introductory section describing in detail the position of the present results relative to earlier and recent works of other authors and comparing the results and methods. Moreover, due to the relative independence of each of the two chapters, each one includes a separate short review on special notions and earlier results. In the beginning of the Chapters 2 and 3 there is a short review on the notations and statements of the anticipative stochastic calculus needed throughout the paper. This anticipative stochastic calculus is developed in detail, e.g., in [33] by D Nualart and E. Pardoux.

1991 Mathematics subject classification: 60H07 (60H10)

Key words: Anticipative stochastic calculus, Skorohod integral, flows of anticipative transformations, Kusuoka's theorem, anticipative Girsanov transformation, enlargement of filtration, anticipative stochastic differential equations with initial condition and with boundary condition, resp., first order stochastic partial differential equations

Acknowledgement

I thank the Professors H. Föllmer, D. Nualart and E. Pardoux for their kind cooperation, which has always been a source of new ideas, impulses and experiences for me. I am very much obliged to Prof. U. Küchler, who has been my head for many years, for his long and permanent efforts to establish an atmosphere allowing his colleagues to concentrate on their research work even under difficult conditions.

Rainer Buckdahn

1 Anticipative stochastic calculus

Let $(\Omega, \mathcal{F}, P)$ denote the standard Wiener space, i.e., Ω is the space of the continuous functions on $[0,1]$ that vanish at 0, $\mathcal{F}$ is the Borel σ-field of subsets of Ω, and P denotes the standard Wiener measure on $(\Omega, \mathcal{F})$. Let $W_t(\omega) = \omega(t)$, $\omega \in \Omega$, $0 \le t \le 1$, be the coordinate process. By H we denote the subspace of all absolutely continuous functions on Ω whose derivative is square integrable, and we endow H with the norm $|h|_H = (\int_0^1 h'(t)^2 dt)^{1/2}$. The norm of the space $L^p([0,1])$ will be denoted by $|.|_p$, i.e. the norm of $L^p(\Omega)$ by $\|.\|_p$, $1 \le p \le \infty$. Consider the space $\mathcal{S}$ of the random variables of the form

$$(1.1) \qquad F = f(W_{t_1}, \ldots, W_{t_n}),$$

where $n \ge 1$; $0 \le t_1, \ldots, t_n \le 1$ and $f \in C_b^\infty(R^n)$. Here $C_b^\infty(R^n)$ denotes the space of C^∞-functions which, together with all their derivatives, are bounded. A random variable of $\mathcal{S}$ will be called a smooth Wiener functional or, shorter, a smooth functional. If F is a smooth functional of the form (1.1), we set for each $0 \le t \le 1$:

$$(1.2) \qquad D_t F = \sum_{i=1}^n \frac{\partial f}{\partial x_i}(W_{t_1}, \ldots, W_{t_n}) I_{[0,t_i)}(t).$$

More generally, we define the k-th derivative of F, for $k \ge 1$, $0 \le s_1, \ldots, s_k \le 1$:

$$D_{s_1,\ldots,s_k}^k F = D_{s_1} \ldots D_{s_k} F.$$

The notation DF stands for the process $(D_t F) = \{D_t F, 0 \le t \le 1\}$ and is called the derivative of F.

Proposition 1.1 *The mapping D is a closable unbounded linear operator from $L^2(\Omega)$ into $L^2([0,1] \times \Omega)$. We identify D with its closed extension, and we denote its domain by $I\!D^{1,2}$.*

More generally, for any $k \ge 1$, $2 \le p < \infty$, we introduce the spaces $I\!D^{k,p}$ as the closure of $\mathcal{S}$ with respect to the norm

$$\|F\|_{k,p} = \|F\|_p + \|(\int_{[0,1]^k} |D_z^k F|^2 dz)^{1/2}\|_p, \quad F \in \mathcal{S}.$$

By $I\!D^{k,\infty}$ and $\widetilde{I\!D^{k,\infty}}$ we denote the restriction of $I\!D^{k,2}$ to those random variables $F \in I\!D^{k,2} \cap L^\infty(\Omega)$ that are of such a form that, the m-parameter process $(D_z^m F) = \{D_z^m F, z \in [0,1]^m\}$ belongs to $L^\infty(\Omega, L^2([0,1]^m))$ and $L^\infty([0,1]^m \times \Omega)$, for all $m = 1, 2, \ldots, k$, respectively.

Received by the editor June 21, 1992.

1

Proposition 1.2 *The operator* $(D, I\!\!D^{1,2})$ *has the following local property: If* $F \in I\!\!D^{1,2}$, *then the process* $(I_{\{F=0\}} D_t F)$ *is equal to zero.*

We now introduce some classes of processes. For $k \geq 1$, $2 \leq p \leq +\infty$, we define

$$I\!\!L^{k,p} = L^p([0,1], I\!\!D^{k,p}) \ (= L^p([0,1] \to I\!\!D^{p,k}; dt)),$$

and

$$\widetilde{I\!\!L^{k,\infty}} = L^\infty([0,1], \widetilde{I\!\!D^{k,\infty}}).$$

Note that

$$\widetilde{I\!\!D^{k,\infty}} \ \subset \ I\!\!D^{\ell,p} \subset I\!\!D^{1,2}$$
$$\widetilde{I\!\!L^{k,\infty}} \ \subset \ I\!\!L^{\ell,p} \subset I\!\!L^{1,2}, \quad \text{for} \quad 1 \leq \ell \leq k, \ 2 \leq p \leq +\infty.$$

If we restrict D to the random variables of $\widetilde{I\!\!D^{1,\infty}}$, we can regard its adjoint δ, called the Skorohod integral, as a closed unbounded linear operator from $L^1([0,1] \times \Omega)$ into $L^1(\Omega)$. That means, the domain of δ is the class of processes $(K_s) \in L^1([0,1] \times \Omega)$ for which there exists an integrable random variable G^K depending only on (K_s) and such that

$$E[G^k \cdot F] = E[\int_0^1 K_s D_s F ds],$$

for all $F \in \mathcal{S}$. The random variable G^K is uniquely determined in $L^1(\Omega)$, it will be denoted by $\delta(K)$ and called the Skorohod integral of (K_s).

The Skorohod integral turns out to be an extension of the classical Itô integral and allows to integrate also nonadapted processes. In addition to the class of adapted square integrable processes, also the processes (K_s) of $I\!\!L^{1,2}$ belong to Dom δ. Moreover, the processes (K_s) of $I\!\!L^{1,2}$ verify $(K_s I_{[0,t]}(s)) \in \text{Dom } \delta$ for each $0 \leq t \leq 1$. For the processes (K_s) having this property we can define the indefinite Skorohod integral $\delta(K_s I_{[0,t]}(s))$ denoted by $\int_0^t K_s dW_s$.

Proposition 1.3 *For each* $0 \leq t \leq 1$, *the Skorohod integral* $\int_0^t K_s dW_s$ *defines a linear continuous mapping of* $I\!\!L^{1,2}$ *into* $L^2(\Omega)$ *which is characterized by the following two properties:*

$$(1.3) \qquad E[\int_0^t K_s dW_s] \ = \ 0,$$

$$E[(\int_0^t K_s dW_s)^2] \ = \ E[\int_0^t K_s^2 ds] + E[\int_0^t \int_0^t D_s K_r D_r K_s dr ds]$$

$$\leq \ \int_0^t \|K_s\|_{1,2}^2 ds.$$

Moreover, this mapping is a local operator, i.e., if $(K_s) \in \mathbb{L}^{1,2}$, then

$$\int_0^t K_s dW_s = 0, \quad a.e. \ on \quad \{\int_0^t K_s^2 ds = 0\}.$$

Proposition 1.4 *If $(K_s) \in \mathbb{L}^{1,2}$, then it holds for any $0 \le t \le 1$ and any increasing sequence of partitions $\Pi_n = \{0 = t_o^n < t_1^n < \ldots < t_n^n = t\}$ with $|\Pi_n| = \max\limits_{1 \le j \le n} (t_j^n - t_{j-1}^n) \to 0 \ (n \to \infty)$ that*

$$\int_0^t K_s dW_s = L^2(\Omega) - \lim_{n \to \infty} \sum_{k=1}^{n-1} E\Big[\frac{1}{t_k^n - t_{k-1}^n} \int_{t_{k-1}^n}^{t_k^n} K_s ds | \mathcal{F}_{[t_k^n, t_{k+1}^n]^c}^W\Big](W_{t_{k+1}^n} - W_{t_k^n}),$$

where $\mathcal{F}_{[t_k^n, t_{k+1}^n]^c}^W$ is the σ-field generated by W_s and $W_t - W_{t_{k+1}^n}, 0 \le s \le t_k^n, t_{k+1}^n \le t \le 1$.

Another possiblity extending the Itô integral to nonadapted integrands (K_s) is given by the extended integral, which was studied, e.g., by M.A. Berger/V.J. Mizel [3]:

Definition 1.5 *A measurable process (K_s) is said to be corrected-integrable if, for all $0 \le t \le 1$ and for any increasing sequence of partitions $\Pi_n = \{0 = t_o^n < t_1^n < \ldots, < t_n^n = t\}$ with $|\Pi_n| = \max\limits_{1 \le j \le n} (t_j^n - t_{j-1}^n) \to 0 \ (n \to \infty)$, there exists the limit of the sequence*

$$\sum_{k=1}^{n-1} \Big(\frac{1}{t_k^n - t_{k-1}^n} \int_{t_{k-1}^n}^{t_k^n} K_s ds\Big)(W_{t_{k+1}^n} - W_{t_k^n})$$

in probability and if this is independent of the special choice of the sequence of partitions. We denote the limit by $\int_0^t K_s \bar{d}W_s$.

Finally, we provide a subset of the domain of the corrected integral:
Let $\mathbb{L}_{C-}^{1,2}$ be the set of those elements (K_s) of $\mathbb{L}^{1,2}$ that satisfy:

   (i) For some $\varepsilon > 0$, the set of processes $\{s \mapsto D_t K_s, (t - \varepsilon)^+ \le s \le t\}_{0 \le t \le 1}$ is equicontinuous with values in $L^2(\Omega)$.

   (ii) $\operatorname*{ess\,sup}\limits_{\substack{0 \le s \le t \le 1, \\ t-s \le \varepsilon}} E[|D_t K_s|^2] < \infty$, for some $\varepsilon > 0$.

Now we can state:

Proposition 1.6 *Let $(K_s) \in \mathbb{L}_{C-}^{1,2}$. Then (K_s) is corrected-integrable, and by the notation*

$$(D_- K)_t = L^2(\Omega) - \lim_{s \to t, s < t} D_t, K_s,$$

the corrected integral is given by

$$\int_0^t K_s \bar{d}W_s = \int_0^t K_s dW_s + \int_0^t (D_- K)_s ds.$$

2 Anticipative Girsanov transformation

Assume that (K_s) is a square integrable process defined on the standard Wiener space $(\Omega, \mathcal{F}, P)$ and possibly not adapted to the natural filtration $(\mathcal{F}_s^W)$ generated by the coordinate process (W_s). The process (K_s) defines a transformation $T : \Omega \to \Omega$ given by $(T\omega)_t = \omega_t + \int_0^t K_s(\omega)ds$, $\omega \in \Omega$, $0 \le t \le 1$. In this chapter we study the problem of the absolute continuity of the image measure $P \circ [T]^{-1}$ and the computation of its density function relative to the Wiener measure P. For the first time, such a transformation without a nonanticipation requirement on (K_s) was studiend by R.H. Cameron and W.T. Martin [17] under strong smoothness assumptions; these results were extended to the abstract Wiener space by L. Gross [25] and H.H. Kuo [28]. R. Ramer [41] introduced an abstract version of the Itô integral, the Ramer-Itô integral, for possibly anticipating processes, which allowed him to present the density $\frac{dQ}{dP}$ for $Q \circ [T]^{-1} = P$ under some conditions on (K_s), among them the continuity of $\omega \in \Omega \mapsto (K_s(\omega)) \in L^2([0,1])$. This was weakened later by S. Kusuoka [29]. The Ramer-Itô integral coincides with the Skorohod integral; so the results of R. Ramer and S. Kusuoka together with the anticipative stochastic calculus provide a useful tool for studying anticipative transformations

$$(2.0.1) \qquad (T\omega)_\bullet = \omega_\bullet + \int_0^\bullet K_s(\omega)ds, \qquad \omega \in \Omega.$$

Nevertheless, it remains quite difficult to compute the density of the transformation, since the formula for the density established by R. Ramer and S. Kusuoka includes the Carleman-Fredholm determinant, for which the known expressions are fairly complicated and allow an explicit computation only in special cases up to now: cf. D.R. Bell [2], A.B. Cruzeiro [18], R. Buckdahn [8], [10] and [11], O. Enchev [21], [22] and A.S. Ustunel/M. Zakai [45].

The common feature of [2], [11], [18], [22] and [45] consists in the embedding of the transformation (2.0.1) into a one-parameter flow

$$(2.0.2) \qquad (T_t\omega)_\bullet = \omega_\bullet + \int_0^\bullet K_{t,s}(\omega)ds, \qquad 0 \le t \le 1,$$

with $T_1 = T$ and $T_o = I$, that of [10] and [21] in the requirement that the $L^2([0,1]^2)$-norm of the Fréchet derivative of the shift process (U_s) of the transformation (2.0.1) should be less than one.

One purpose of this chapter is to study flows (2.0.2) of transformations with a shift process

$$(2.0.3) \qquad K_{t,s}(\omega) = I_{[0,t]}(s)\sigma_s(T_s\omega), \qquad 0 \le s,t \le 1,$$

4

where (σ_s) is a possibly anticipating process. This problem was first studied in [5], with the motivation of developing a method for solving linear and quasilinear stochastic differential equations with anticipation, cf. [4] and [5].

Meanwhile, lots of generalizations of [5] and [11] have been developed: While [5] and [11] consider the Fréchet differentiable process (σ_s) on the standard Wiener space, the same subject is studied in [13], [14] over the abstract Wiener space. In both cases the process (σ_s) as well as the Fréchet derivative of (σ_s) are supposed to be bounded. In a recent paper [22] O. Enchev has shown that the same results remain true if the requirement of the boundedness of (σ_s) is replaced by its square integrability. Another way of generalization has been found by A.S. Ustunel/M. Zakai [45] who has replaced the boundedness assumption on (σ_s) by a requirement on the integrability of a certain moment of the exponent of the norm of (σ_s) in $L^2([0,1])$. This requirement allows to consider conditions on the Fréchet derivative of (σ_s) that are weaker than those of Enchev.

Another purpose of this chapter consists in studying the transformations (2.0.1) defined by a possibly anticipating shift process. For such transformations D. Nualart and M. Zakai have established a Girsanov-type theorem under the rather general assumption of a Fréchet differentiable weakly one-to-one process (K_s). However, here it remains unclear which processes (K_s) are weakly one-to-one. A class of such processes, namely the square integrable processes (K_s) for which the $L^2([0,1]^2)$-norm of the Fréchet derivative is less than one, has been studied in a recent paper [21] by O. Enchev. Let I be the identity operator on $L^2([0,1])$. Under the Gohberg-Krein factorization

$$(I + DK(\omega))^{-1} = (I + V^+(\omega))(I + V^-(\omega)),$$
$$(V_{s,t}^+(\omega)) \cdot (V_{s,t}^-(\omega)) \in L^2([0,1]^2) \quad \text{s.t.} \quad V_{s,t}^+(\omega) = V_{t,s}^-(\omega) = 0, \quad \text{if} \quad s < t,$$

and with the notation

$$L = \exp\left\{ -\int_0^1 K_s dW_s - \frac{1}{2} \int_0^1 K_s^2 ds - \int_0^1 \int_0^s D_r K_s V_{s,r}^+ dr ds \right\},$$

Enchev has proved that the Girsanov-type statement holds true: If $E[L] = 1$, then

$$E[F(T)L] = E[F], \quad \text{for all} \quad F \in L^\infty(\Omega).$$

The earlier paper [10] shows that, if (K_s) satisfies some Novikov-type condition and has a Fréchet derivative whose $L^2([0,1]^2)$-norm is bounded by a constant less than one, then $E[L] = 1$ and, moreover, the Volterra kernels $(V_{s,t}^+)$, $(V_{s,t}^-)$ are computed explicitly.

The chapter is organized as follows: The first section presents a short review on the stochastic calculus with transformations. The statements are given without proofs, for details we refer to [8], [10], [24] and [29] for instance. Section 2.2 is devoted to families of transformations (2.0.2) embedded to one-parameter flow (2.0.3) whose possibly anticipating shift process (σ_s) is Fréchet differentiable.

The approach we will use here was originally developed for a common work with

H. Föllmer [15] and consists in passing to a product space on which (σ_s), first assumed to be a smooth step process, is nonanticipating. Once having the main result for such smooth step processes (σ_s), it is not hard to generalize this result to Fréchet differentiable processes, and also to not necessarily Fréchet differentiable processes (σ_s) adapted to some enlargement of the canonical filtration $(\mathcal{F}_s^W)$. This will be done in Section 2.3. In the Sections 2.4 and 2.5 transformations of the form (2.0.1) with a Fréchet differentiable shift process (K_s) are considered, in 2.4 with the requirement that the $L^2([0,1]^2)$-norm of the Fréchet derivative of (K_s) is less than one, in 2.5 with the assumption of adaptedness to some enlargement on $(\mathcal{F}_s^W)$. While Section 2.4 bases mainly on [10] and [11], Section 2.5 has its origin in [15].

2.1   Stochastic calculus for transformations. Short Review

We consider mappings $T : \Omega \to \Omega$ of the form

$$T\omega = \omega + \int\limits_0^{\bullet} K_s(\omega)ds, \quad \omega \in \Omega,$$

where the shift process (K_s) is assumed to belong to $L^2([0,1] \times \Omega)$. Such mappings are called transformations for short.

Definition 2.1.1 *We call the transformation* T

  (i) absolutely continuous if the measure $P \circ [T]^{-1}$ *with*

$$P \circ [T]^{-1}(B) = P\{\omega : T\omega \in B\}, \quad B \in \mathcal{F},$$

  is absolutely continuous relative to P, *and*

  (ii) invertible with an inverse transformation $A : \Omega \to \Omega$ *if, for some versions of* T *and* A *also denoted by* T *and* A, *respectively, it holds*

$$T(A\omega) = A(T\omega) = \omega, \quad \omega \in \Omega.$$

Note that, the measures P, $P \circ [T]^{-1}$ and $P \circ [A]^{-1}$ are equivalent for any absolutely continuous and invertible transformation T.

Before describing properties of absolutely continuous transformations we will present a statement that allows us to reduce the study of general transformations to those with a smooth shift process.

Proposition 2.1.2 *Let* $F \in \mathbb{D}^{1,2}$. *Then, for any* $\varepsilon > 0$, *there exists a sequence of functionals* $(F^n) \subset \mathcal{S}$ *with* $\|F - F^n\|_{1,2} \to 0$ $(n \to \infty)$ *such that*

  (i) $\|F^n\|_\infty \le \|F\|_\infty$,

  (ii) $\|(\int\limits_0^1 |D_s F^n|^2 ds)^{1/2}\|_\infty \le \varepsilon + \|(\int\limits_0^1 |D_s F|^2 ds)^{1/2}\|_\infty$,
      and

(iii) $\|DF^n\|_{L^\infty([0,1]\times\Omega)} \leq \varepsilon + \|DF\|_{L^\infty([0,1]\times\Omega)}$, $n = 1, 2, 3, \ldots$.

For the proof we refer to Proposition 2.5 [10]. In Proposition 2.1.3 (Propositions 2.6 [10] and 2.14 [8]) the statement 2.1.2 will be extended to processes of $\mathbb{L}^{1,2}$, where the role of the smooth Wiener functionals of $\mathcal{S}$ in 2.1.2 will be replaced by smooth step processes.

A process (K_s) is called a smooth step process if there exists a partition $0 = t_o < t_1 < \ldots < t_n = 1$ $(n \geq 1)$ and random variables $F_i \in \mathcal{S}$, $i = 1, 2, \ldots, n$, such that

$$(2.1.1) \qquad K_s(\omega) = \sum_{i=1}^{n} F_i(\omega) I_{[t_{i-1}, t_i)}(s), \quad (s, \omega) \in [0, 1] \times \Omega.$$

The set of the smooth step processes will be denoted by $\mathbb{L}^{\mathcal{S}}$.

Proposition 2.1.3 *Let $(K_s) \in \mathbb{L}^{1,2}$. Then, for any $\varepsilon > 0$, there is a sequence $((K_s^n))$ of smooth step processes with*

$$\int_0^1 \|K_s^n - K_s\|_{1,2}^2 ds \to 0 \quad (n \to \infty)$$

such that

(i) $\|(\int_0^1 |K_s^n|^2 ds)^{1/2}\|_\infty \leq \|(\int_0^1 |K_s|^2 ds)^{1/2}\|_\infty$,

(ii) $\|(\int_0^1 \int_0^1 |D_s K_t^n|^2 ds dt)^{1/2}\|_\infty \leq \varepsilon + \|(\int_0^1 \int_0^1 |D_s K_t|^2 ds dt)^{1/2}\|_\infty$,
  and

(iii) $(\int_0^1 \|\int_0^1 |D_s K_t^n|^2 ds\|_\infty dt)^{1/2} \leq \varepsilon + (\int_0^1 \|\int_0^1 |D_s K_t|^2 ds\|_\infty dt)^{1/2}$, $n = 1, 2, 3, \ldots$,

As a direct consequence of the Propositions 2.1.2 and 2.1.3 we obtain the following statements 2.1.4 - 2.1.6.

Proposition 2.1.4 *(Prop. 2.7. [10]) Let T^1, T^2 be two transformations. Assume that either*

(i) $F \in \mathcal{S}$, *or*

(ii) $F \in \mathbb{D}^{1,2}$ *and T^1, T^2 are absolutely continuous.*

Then

$$|F(T^1\omega) - F(T^2\omega)| \leq \|(\int_0^1 |D_s F|^2 ds)^{1/2}\|_\infty \cdot |T^1\omega - T^2\omega|_H, \quad a.e.$$

Proposition 2.1.5 *(Prop. 2.8 [10]) Let $F \in \mathbb{D}^{1,2}$, and let T be a transformation with shift process $(K_s) \in \mathbb{L}^{1,2}$. Assume that either*

(i) $F \in \mathcal{S}$, or

(ii) T is absolutely continuous, $F(T\omega) \in L^2(\Omega)$ and $\|(\int_0^1 |D_s F|^2 ds)^{1/2}\|_\infty < +\infty$.

Then, $F(T\omega) \in \mathbb{D}^{1,2}$, and

$$D_t[F(T\omega)] = (D_t F)(T\omega) + \int_0^1 (D_s F)(T\omega) D_t K_s(\omega) ds, \quad a.e.$$

In case $F \in \mathcal{S}$, $(K_s) \in \mathbb{L}^{\mathcal{S}}$, this equation holds for all $0 \le t \le 1$.

The Propositions 2.1.4 and 2.1.5 remain true if we consider a process $(\sigma_s) \in \mathbb{L}^{\mathcal{S}}$ ($(\sigma_s) \in \mathbb{L}^{1,2}$) instead of a random variable $F \in \mathcal{S}$ ($F \in \mathbb{D}^{1,2}$, respectively): In Proposition 2.1.4 we obtain

$$(\int_0^1 |\sigma_s(T^1\omega) - \sigma_s(T^2\omega)|^2 ds)^{1/2} \le \|(\int_0^1 \int_0^1 |D_r\sigma_s|^2 dr ds)^{1/2}\|_\infty \cdot |T^1\omega - T^2\omega|_H, \quad a.e.,$$

and, under the requirements that either $(\sigma_s) \in \mathbb{L}^{\mathcal{S}}$ or $(\sigma_s) \in \mathbb{L}^{1,2}$, $T : \Omega \to \Omega$ is absolutely continuous, $(\sigma_s(T)) \in L^2([0,1] \times \Omega)$ and

$$\|(\int_0^1 \int_0^1 |D_r\sigma_s|^2 dr ds)^{1/2}\|_\infty < \infty,$$

Proposition 2.1.5 yields

$$(\sigma_s(T)) \in \mathbb{L}^{1,2} \quad \text{and}$$

$$D_t[\sigma_s(T\omega)] = (D_t\sigma_s)(T\omega) + \int_0^1 (D_r\sigma_s)(T\omega) D_t K_r(\omega) dr, \quad a.e.$$

The rule for changing the order of differentiation and transformation allows us to deduce a rule for changing the order of integration and transformation of a process (Prop. 2.19 of [8]).

Proposition 2.1.6 *Let $(\sigma_s) \in \mathbb{L}^{1,2}$, and let T be a transformation with shift process $(K_s) \in \mathbb{L}^{1,2}$. Assume that either*

(i) (σ_s) *is a smooth step process, or*

(ii) T *is absolutely continuous, $(\sigma_s(T\omega)) \in L^2([0,1] \times \Omega)$ and*

$$\|(\int_0^1 \int_0^1 |D_r\sigma_s|^2 dr ds)^{1/2}\|_\infty < +\infty.$$

Then, $(\sigma_s(T\omega)) \in \mathbb{L}^{1,2}$, and

$$\int_0^1 \sigma_s(T\omega)dW_s = (\int_0^1 \sigma_s dW_s)(T\omega) - \int_0^1 \sigma_s(T\omega)K_s(\omega)ds$$

$$- \int_0^1 \int_0^1 (D_r\sigma_s)(T\omega)D_sK_r(\omega)drds, \quad a.e.$$

Finally we recall statements on the convergence of sequences of transformations. The following proposition is due to Gihman and Skorohod, Theorems 2 and 3 [24].

Proposition 2.1.7 *Let $(T^n\omega = \omega + \int_0^\bullet K_s^n(\omega)ds)$ be a sequence of absolutely continuous transformations such that*

(i) the sequence of processes $((K_s^n))$ converges in $L^2([0,1] \times \Omega)$ to some process (K_s), and

(ii) the sequence of densities $(L^n = \dfrac{dP \circ [T^n]^{-1}}{dP})$ is uniformly integrable.

Then the transformation

$$T\omega = \omega + \int_0^\bullet K_s(\omega)ds$$

is absolutely continuous, and the density L of T is the limit of (L^n) in the topology $\sigma(L^1, L^\infty)$.

Due to Proposition 2.1.7 we can state:

Proposition 2.1.8 *(Prop. 2.10 [10]) Let $(T^n\omega = \omega + \int_0^\bullet K_s^n(\omega)ds)$ be a sequence of absolutely continuous transformations such that*

(i) the sequence $((K_s^n))$ converges in $L^2([0,1] \times \Omega)$ to some (K_s), and

(ii) $(L^n = \dfrac{dP \circ [T^n]^{-1}}{dP})$ is uniformly integrable.

Then, with the notation $T\omega = \omega + \int_0^\bullet K_s(\omega)ds$, the convergence in probability of any sequence (F^n) to some random variable F implies

$$F(T\omega) = \lim_{n\to\infty} F^n(T^n\omega), \quad in\ probability.$$

Remark: Proposition 2.1.8 can be modified as follows: If $(T^n\omega = \omega + \int_0^\bullet K_s^n(\omega)ds)$ is a sequence of absolutely continuous transformations such that

(i) the assumptions of Proposition 2.1.8 are satisfied, and

(ii') the sequence of the densities $\mathcal{L}^n$ with

$$E[F(T^n)\mathcal{L}^n] = E[F], \quad \text{for all} \quad F \in L^\infty(\Omega),$$

converges in $L^1(\Omega, \mathcal{F}, P)$ and the sequence of the inverses $((\mathcal{L}^n)^{-1})$ is uniformly integrable.

Then, with the notation $T\omega = \omega + \int_0^\bullet K_s(\omega)ds$, the transformation T is absolutely continuous, and the convergence in probability of any sequence (F^n) to some random variable F implies

$$F(T\omega) = \lim_{n\to\infty} F^n(T^n\omega), \quad \text{in probability.}$$

Since the Propositions 2.1.2, 2.1.7 and 2.1.8 permit to reduce the study of the transformation $T\omega = \omega + \int_0^\bullet K_s(\omega)ds$, $K \in \mathbb{L}^{1,2}$, to transformations with shift in $\mathbb{L}^S$, the extended Girsanov theorem of Kusuoka (Theorem 6.4 [29]) can be used as an important tool in the analysis of transformations (like, e.g., in [10] and [11]):

Proposition 2.1.9 Let $T : \Omega \to \Omega$ be a transformation of the form $T\omega = \omega + \int_0^\bullet K_s(\omega)ds$, where $(K_s) \in L^2([0,1] \times \Omega)$, and suppose that the following conditions are satisfied for some version of (K_s):

(i) T is bijective.

(ii) For all $\omega \in \Omega$, there is an element $DK(\omega) \in L^2([0,1]^2)$ such that

(1) $|K_\bullet(\omega + \int_0^\bullet h_s ds) - K_\bullet(\omega) - \int_0^1 D_s K_\bullet(\omega)h_s ds|_2 = o(|h|_2)$, as $|h|_2 \to 0$, $h \in L^2([0,1])$.

(2) $h \mapsto DK(\omega + \int_0^\bullet h_s ds)$ is continuous from $L^2([0,1])$ into $L^2([0,1]^2)$.

(3) The mapping $I + DK(\omega) : h \mapsto h + \int_0^1 D_s K_\bullet(\omega)h_s ds$ from $L^2([0,1])$ into itself is invertible.

Then the transformation T is absolutely continuous, and its inverse transformation A has the density

$$\frac{dP \circ [A]^{-1}}{dP} = |d_c(-DK)| \exp\{-\int_0^1 K_s dW_s - \frac{1}{2}\int_0^1 K_s^2 ds\},$$

where $d_c(-DK)$ denotes the Carleman-Fredholm determinant of the Hilbert-Schmidt operator $-DK : h \to -\int_0^1 D_s K_\bullet h_s ds$.

The Carleman-Fredholm determinant of a Hilbert-Schmidt operator B from $L^2([0,1])$ into itself is defined by the product expansion

$$d_c(B) = \prod_j (1 - \lambda_j)e^{\lambda_j}.$$

Here the λ_j's are the nonzero eigenvalues of B counted with their multiplicities. In particular, if the operator B is nuclear, then

$$d_c(B) = \det(I - B)\exp\{\text{trace } B\}.$$

Since $d_c(.) : L^2([0,1]^2) \to R^1$ is continuous, the Carleman-Fredholm determinant $d_c(B)$ can be computed for non-nuclear B by approximating B by nuclear operators.

2.2 Transformations given by a smooth flow

In this section we want to study a family of transformations $\{T_t, 0 \le t \le 1\}$ of Ω into itself which verifies the equation

$$(2.2.1) \qquad T_t\omega = \omega + \int_0^{t\wedge .} \sigma_s(T_s\omega)ds,$$

for a.e. $\omega \in \Omega$ and for each $0 \le t \le 1$, where we suppose the process (σ_s) to belong to $L^2([0,1], I\!D^{1,\infty})$. The main result of this section is the following:

Theorem 2.2.1 *Assume that (σ_s) belongs to $L^2([0,1], I\!D^{1,\infty})$. Then there exists a unique family $\{T_t, 0 \le t \le 1\}$ of absolutely continuous transformations $T_t : \Omega \to \Omega$ that satisfies the equation (2.2.1). Moreover, for each $0 \le t \le 1$, the transformation $T_t : \Omega \to \Omega$ is invertible, and its inverse transformation A_t has the density*

$$(2.2.2) \qquad \frac{dP \circ [A_t]^{-1}}{dP} = \exp\{-\int_0^t \sigma_s(T_s)dW_s - \frac{1}{2}\int_0^t \sigma_s(T_s)^2 ds$$
$$- \int_0^t \int_0^s (D_r\sigma_s)(T_s)D_s[\sigma_r(T_r)]drds\}.$$

Here, together with (σ_s), the process $(\sigma_s(T_s))$ is in $L^2([0,1], I\!D^{1,\infty})$.

The derivation of Theorem 2.2.1 requires first to study equation (2.2.1) for a smooth step process $(\sigma_s) \in I\!L^S$. For this, note that if $(\sigma_s) \in I\!L^S$, then we can find a natural n, functions $f_1, \ldots, f_n \in C_b^\infty(R^n)$, and a partition $0 = t_o < t_1 < \ldots < t_n = 1$ of the interval $[0,1]$ into subintervals $\Delta_j = [t_{j-1}, t_j)$ with the length $|\Delta_j|$ such that, with the notations

$$e_j(s) = \frac{1}{\sqrt{|\Delta_j|}}I_{\Delta_j}(s), \qquad \omega(e_j) = \frac{1}{\sqrt{|\Delta_j|}}(\omega(t_j) - \omega(t_{j-1}))$$

and

$$f_s(x) = \sum_{j=1}^n f_j(x)e_j(s), \quad (s, x) \in [0,1] \times R^n,$$

(σ_s) has the representation

$$(2.2.3) \qquad \sigma_s(\omega) = f_s(\omega(e_1), \ldots, \omega(e_n)), \quad (s, \omega) \in [0, 1] \times \Omega.$$

Lemma 2.2.2 *If $(\sigma_s) \in \mathbb{L}^S$, then there is a unique family of transformations $\{T_t, 0 \leq t \leq 1\}$ satisfying equation (2.2.1). Moreover, for any $0 \leq t \leq 1$, the transformation $T_t : \Omega \to \Omega$ is invertible, its inverse A_t is given by $A_t = A_{o,t}$, where the family of transformations $A_{s,t} : \Omega \to \Omega$, $0 \leq s \leq t$, is the unique solution of the equation*

$$(2.2.4) \qquad A_{s,t}\omega = \omega - \int_{s\wedge.}^{t\wedge.} \sigma_r(A_{r,t}\omega)dr.$$

The family $\{A_{s,t}, 0 \leq s \leq t \leq 1\}$ forms a semi-group, in particular, it holds $A_{s,t} = T_s \circ A_t$, $0 \leq s \leq t \leq 1$.

Proof: The correctness of the statement follows immediately from the special form (2.2.3) of (σ_s), which allows us to find a general constant $C > 0$ such that

$$\begin{aligned} |\sigma_s(\omega)| &\leq C, \\ |\sigma_s(\omega') - \sigma_s(\omega'')| &\leq C \sup_{0 \leq t \leq 1} |\omega'(t) - \omega''(t)|, \end{aligned}$$

for all $0 \leq s \leq 1$, $\omega, \omega', \omega'' \in \Omega$. Hence, we can use standard arguments from analysis that provide the desired statement.

For the transformations $A_t, T_t : \Omega \to \Omega$ introduced in Lemma 2.2.2 we can state:

Proposition 2.2.3 *For any $0 \leq t \leq 1$, the transformations A_t and T_t of Ω into themselves are absolutely continuous, their densities are given by*

$$(2.2.5)$$

$$\frac{dP \circ [A_t]^{-1}}{dP} = \exp\{-\int_0^t \sigma_r(T_r)\bar{d}W_r - \frac{1}{2}\int_0^t \sigma_r(T_r)^2 dr - \int_0^t (D_r\sigma_r)(T_r)dr\}, \ 0 \leq t \leq 1,$$

and

$$\frac{dP \circ [T_t]^{-1}}{dP} = \exp\{\int_0^t \sigma_s(A_{s,t})\bar{d}W_s - \frac{1}{2}\int_0^t \sigma_s(A_{s,t})^2 ds - \int_0^t (D_s\sigma_s)(A_{s,t})ds\}, \ 0 \leq t \leq 1.$$

In particular, the corrected integrals in (2.2.5) exist, indeed.

Proof: We use the notations introduced above. Additionally, let $Y = (Y_{s,t}(x) = (Y^1_{s,t}(x), \ldots, Y^n_{s,t}(x)))) : \{(s,t) \in [0,1]^2 : s \leq t\} \times R^n \to R^n$ be the solution of the equation

$$(2.2.6) \quad Y^i_{s,t}(x) = x_i - \int_s^t f_r(Y_{r,t}(x))e_i(r)dr, \qquad 0 \leq s \leq t \leq 1, \ i = 1, 2, \ldots, n,$$

$$x = (x_1, \ldots, x_n) \in R^n.$$

Note that, for all $0 \le s \le t \le 1$, $Y_{s,t} : R^n \to R^n$ is a bijection with Jacobian determinant

$$\det(Y'_{s,t}(x)) = \exp\{-\int_s^t \sum_{i=1}^n (\frac{\partial}{\partial x_i} f_r)(Y_{r,t}(x))e_i(r)dr\}.$$

Use (2.2.6) to define the nonanticipating transformation $A_{s,t}(.,x) : \Omega \to \Omega$,

$$A_{s,t}(\omega, x) = \omega - \int_{s\wedge.}^{t\wedge.} f_r(Y_{r,t}(x))dr, \quad 0 \le s \le t \le 1, \ x \in R^n,$$

and set

$$A_t(\omega, x) = A_{o,t}(\omega, x).$$

Then, obviously,

$$A_t\omega = A_t(\omega, \bar{\omega}), \qquad 0 \le t \le 1, \ \omega \in \Omega,$$

where $\bar{\omega}$ denotes the vector $(\omega(e_1), \dots, \omega(e_n))$. Since the shift of $A_t(.,x)$ is deterministic, $A_t(.,x) : \Omega \to \Omega$ is invertible, its inverse $T_t(.,x)$ being of the form

$$T_t(\omega, x) = \omega + \int_0^{t\wedge.} f_r(Y_{r,t}(x))dr, \ 0 \le t \le 1, \ x \in R^n.$$

Recall that the nonanticipative transformations $A_{s,t}(.,x)$ and $T_t(.,x) : \Omega \to \Omega$ induce image measures that are equivalent to the Wiener measure P, in particular

$$(2.2.7) \qquad L_t(x) \ = \ \frac{dP \circ [T_t(.,x)]^{-1}}{dP} =$$

$$= \ \exp\{\int_0^t f_r(Y_{r,t}(x))dW_r - \frac{1}{2}\int_0^t f_r(Y_{r,t}(x))^2 dr\}.$$

Taking into account the existence of the derivative $\frac{d}{dt}Y_{s,t}(x)$, which is bounded relative to (s,t,x), $0 \le s \le t \le 1$, $x \in R^n$, we see that

$$\sup_{x \in R^n} E[\sup_{0 \le t \le 1} \{|L_t(x)| + \sum_{i=1}^n |\frac{\partial}{\partial x_i}L_t(x)|\}^p] < \infty, \text{ for all } 1 < p < \infty.$$

Since $\det(Y'_{o,t}(x))$ is bounded on $[0,1] \times R^n$, the arguments used for the proof of the Lemmata 2.1. and 2.2 of [38] by D. Ocone/E. Pardoux provide the existence of a random variable $\zeta \in \bigcap_{p>1} L^p(\Omega)$ such that

$$0 \ \le \ L_t(x) \det(Y'_{o,t}(x)) \le \zeta(1 + |x|),$$
$$\text{for all } (t,x) \in [0,1] \times R^n, \text{ a.e.}$$

Hence, in virtue of the obvious continuity of the function
$t \mapsto L_t(\omega, x) \det(Y'_{o,t}(x))F(A_t(\omega, x))$, for a.e. $\omega \in \Omega$, the dominated convergence theorem implies

$$(2.2.8) \ \ E[L_t(\bar{\omega}) \det(Y_{o,t}(\bar{\omega}))F(A_t\omega)] =$$

$$= \ \lim_{\varepsilon \downarrow 0} E[\int_{R^n} L_t(\bar{\omega} + \varepsilon x) \det(Y_{o,t}(\bar{\omega} + \varepsilon x))F(A_t(\bar{\omega} + \varepsilon x))\mathcal{N}(0, I_n)(dx)].$$

However, this gives the possibility to eliminate the random vector $\bar{\omega}$ in the transformation $A(\bar{\omega} + \varepsilon x)$ and to pass to the nonanticipative transformation $A(x)$. For this, fix any $0 < \varepsilon < 1$ and consider the corresponding expression on the right-hand side of (2.2.8). With the notation

$$\varrho_\varepsilon(x) = \frac{\mathcal{N}(0, \varepsilon^2 I_n)(dx)}{dx} = \frac{1}{(2\pi)^{n/2}\varepsilon^n} \exp\{-\frac{1}{2}|\frac{x}{\varepsilon}|^2\}, \ x \in R^n,$$

Fubini's theorem and the Girsanov transformation theorem for deterministic shifts yield

$$(2.2.9) \qquad E[\int\limits_{R^n} L_t(\bar{\omega} + \varepsilon x) \det(Y'_{o,t}(\bar{\omega} + \varepsilon x)) F(A_t(\bar{\omega} + \varepsilon x)) \mathcal{N}(0, I_n)(dx)]$$

$$= \int\limits_{R^n} E[L_t(x) F(A_t(x)) \varrho_\varepsilon(x - \bar{\omega})] \det(Y'_{o,t}(x)) dx$$

$$= \int\limits_{R^n} E[F \varrho_\varepsilon(x - \overline{T_t(\omega, x)})] \det(Y'_{o,t}(x)) dx.$$

Here $\overline{T_t(\omega, x)}$ denotes the vector $(T_t(\omega, x)(e_1), \ldots, T_t(\omega, x)(e_n))$. Obviously,

$$x - \overline{T_t(\omega, x)} = Y_{o,t}(x) - \bar{\omega},$$

i.e., (2.2.9) takes the form

$$(2.2.10) \qquad E[\int\limits_{R^n} L_t(\bar{\omega} + \varepsilon x) \det(Y'_{o,t}(\bar{\omega} + \varepsilon x)) F(A_t(\bar{\omega} + \varepsilon x)) \mathcal{N}(0, I_n)(dx)]$$

$$= E[F \int\limits_{R^n} \det(Y'_{o,t}(x)) \varrho_\varepsilon(Y_{o,t}(x) - \bar{\omega}) dx].$$

Taking into account that the mapping $Y_{o,t} : R^n \to R^n$ is invertible, we can substitute $y = Y_{o,t}(x)$ on the right-hand side of (2.2.10) and obtain

$$E[\int\limits_{R^n} L_t(\bar{\omega} + \varepsilon x) \det(Y'_{o,t}(\bar{\omega} + \varepsilon x)) F(A_t(\bar{\omega} + \varepsilon x)) \mathcal{N}(0, I_n)(dx)]$$

$$= E[F], \quad \text{for any } 0 < \varepsilon \leq 1, \ F \in \mathcal{S}.$$

Therefore, (2.2.8) yields

$$E[L_t(\bar{\omega}) \det(Y'_{o,t}(\bar{\omega})) F(A_t\omega)] = E[F], \quad \text{for all } F \in \mathcal{S}.$$

This implies the equivalence of $P \circ [A_t]^{-1}$ and $P \circ [T_t]^{-1}$ to P, and

$$L_t = \frac{dP \circ [T_t]^{-1}}{dP} = L_t(\bar{\omega}) \det(Y'_{o,t}(\bar{\omega})).$$

Clearly, the corrected integral $\int\limits_0^t f_s(Y_{s,t}(\bar{\omega})) \bar{d}W_s$ exists and coincides with $I_t(x) = \int\limits_0^t f_s(Y_{s,t}(x)) dW_s$ at $x = \bar{\omega}$. In order to check this, it suffices to show that for any increasing sequence of partitions $\Pi_m = \{0 = t_o^m < t_1^m < \cdots < t_m^m = t\}$ with

$$|\Pi_m| = \max_{1 \leq j \leq m} (t_j^m - t_{j-1}^m) \to 0, \quad m \to \infty,$$

the sequence

$$
\mathcal{E}_m(x) = \sum_{j=1}^{m-1} \left(\frac{1}{t_j^m - t_{j-1}^m} \int_{t_{j-1}^m}^{t_j^m} f_s(Y_{s,t}(x)) ds \right) \cdot (W_{t_{j+1}^m} - W_{t_j^m}),
$$

$$
m = 1, 2, 3, \ldots, \ x \in R^n,
$$

is such a form that $I_t(\bar{\omega}) = \lim\limits_{m\to\infty} \mathcal{E}_m(\bar{\omega})$ holds in probability under each law $P^x = P(.|\bar{\omega} = x)$, $x \in R^n$. However, this follows from

$$
P^x\{|I_t(\bar{\omega}) - \mathcal{E}_m(\bar{\omega})| > \varepsilon\} \le P\{|I_t(x) - \mathcal{E}_m(x)| > \frac{\varepsilon}{2}\}
$$

$$
+ \ P\{\sum_{k=1}^{n} |W(e_k) - x_k| \, \Big| \int_0^t (f_s(Y_{s,t}(x)) e_k(s) ds
$$

$$
- \sum_{j=1}^{m-1} \left(\frac{1}{t_j^m - t_{j-1}^m} \int_{t_{j-1}^m}^{t_j^m} f_s(Y_{s,t}(x)) ds \right) \int_{t_j^m}^{t_{j+1}^m} e_k(s) ds \Big| > \frac{\varepsilon}{2}\}
$$

$$
\to 0, \text{ as } m \to \infty, \text{ for all } x = (x_1, \ldots, x_n) \in R^n \text{ and for all } \varepsilon > 0.
$$

Hence, $\int_0^t f_s(Y_{s,t}(\bar{\omega})) \bar{d}W_s = I_t(\bar{\omega})$, and so we conclude from (2.2.6) and (2.2.7)

$$
L_t = \exp\{ \int_0^t f_s(Y_{s,t}(\bar{\omega})) \bar{d}W_s - \frac{1}{2} \int_0^t f_s(Y_{s,t}(\bar{\omega}))^2 ds - \int_0^t \sum_{i=1}^{n} (\frac{\partial}{\partial x_i} f_s)(Y_{s,t}(\bar{\omega})) e_i(s) ds \}.
$$

Due to (2.2.3) and the definitions of $A_{s,t}$ and $Y_{s,t}(x)$, respectively, we have

$$
f_s(Y_{s,t}(\bar{\omega})) = \sigma_s(A_{s,t}),
$$

$$
\sum_{i=1}^{n} (\frac{\partial}{\partial x_i} f_s)(Y_{s,t}(\bar{\omega})) e_i(s) = (D_s \sigma_s)(A_{s,t}).
$$

Therefore, formula (2.2.5) for the density of T_t is true. It remains to compute the density $\mathcal{L}_t$ of A_t. For this note that, for any $F \in \mathcal{S}$,

$$
E[F(A_t)] = E[(F L_t(T_t)^{-1}) \circ A_t \cdot L_t] = E[F L_t(T_t)^{-1}].
$$

Consequently,

$$
\mathcal{L}_t = \frac{dP \circ [A_t]^{-1}}{dP} = L_t(T_t)^{-1}.
$$

Since $A_{s,t} \circ T_t = T_s$, and

$$
(\int_0^t \sigma_s(A_{s,t}) \bar{d}W_s) \circ T_t = \int_0^t \sigma_s(T_s) \bar{d}W_s + \int_0^t \sigma_s(T_s)^2 ds,
$$

the random variable $L_t(T_t)^{-1}$ is nothing else but the right-hand side of the first line of (2.2.5). This completes the proof.

Remark: Taking into account the decomposition $A_{s,t} = T_s \circ A_t$, $0 \leq s \leq t \leq 1$, we can conclude from Lemma 2.2.3 that also the transformation $A_{s,t}$ is absolutely continuous with respect to P. Set $L_{s,t} = L_s(A_{s,t})^{-1}L_t$.
Then, for all $F \in L^\infty(\Omega)$,

$$E[F(A_{s,t})L_{s,t}] (= E[F(A_{s,t})L_s(A_{s,t})^{-1}L_t] = E[F(T_s)L_s(T_s)^{-1}]) = E[F].$$

Moreover, (2.2.5) allows us to compute $L_{s,t} = L_s(T_s)^{-1} \circ A_t \cdot L_t$: If we consider that

$$\left(\int_0^s \sigma_r(A_{r,s})\bar{d}W_r\right) \circ A_{s,t} = \int_0^s \sigma_r(A_{r,t})\bar{d}W_r,$$

then

$$(2.2.11) \quad L_{s,t} \;=\; \exp\{\int_s^t \sigma_r(A_{r,t})\bar{d}W_r - \frac{1}{2}\int_s^t \sigma_r(A_{r,t})^2 dr - \int_s^t (D_r\sigma_r)(A_{r,t})dr\},$$

$$0 \leq s \leq t \leq 1.$$

For later purposes we need the representation of the density $\mathcal{L}_t$ of A_t in the form (2.2.2) and the corresponding analogon for the density of T_t. This requires to compute the occurring corrected integrals by Skorohod integrals. For this purpose, we first state:

Lemma 2.2.4 *If $F \in \mathcal{S}$, and $\{T_t, 0 \leq t \leq 1\}$ and $\{A_{s,t}, 0 \leq s \leq t \leq 1\}$ are the families of transformations associated with $(\sigma_s) \in \mathbb{L}^S$ by (2.2.1) and (2.2.4), respectively, then the processes $(F(T_t))$ and $(F(A_{s,t})I_{[0,t]}(s))$ belong to $\widetilde{\mathbb{L}^{1,\infty}}$, $F(T_t)$ as well as $F(A_{s,t})$ are in $\mathcal{S}$ and, moreover, the mappings $t \mapsto F(T_t)$ and $s \mapsto F(A_{s,t})$ are pathwise differentiable,*

$$\frac{d}{dt}F(T_t) \;=\; \sigma_t(T_t)(D_tF)(T_t), \quad 0 \leq t \leq 1,$$
$$\frac{d}{ds}F(A_{s,t}) \;=\; \sigma_s(A_{s,t})(D_sF)(A_{s,t}), \quad 0 \leq s \leq t \leq 1.$$

Proof: Suppose that (σ_s) has the representation (2.2.3). Without loss of generality we can assume that

$$F = \varphi(W(e_1), \ldots, W(e_n)), \quad \varphi \in C_b^\infty(R^n).$$

In the other case we have to choose another representation (2.2.3). Then, using the notations of the proof of Proposition 2.2.3, we have

$$F(A_{s,t}) = \varphi(Y_{s,t}(\bar{\omega})).$$

This immediately implies the statement for $(F(A_{s,t})I_{[0,t]}(s))$, the statement for $(F(T_t))$ can be proved analogously.

Lemma 2.2.5 *Let* $(\sigma_s) \in \mathbb{L}^S$. *With the notations introduced above it holds*

(2.2.12)
$$\int_0^t \sigma_s(T_s)\bar{d}W_s = \int_0^t \sigma_s(T_s)dW_s + \int_0^t D_s[\sigma_s(T_s)]ds,$$

and

(2.2.13)
$$\int_0^t \sigma_s(A_{s,t})\bar{d}W_s = \int_0^t \sigma_s(A_{s,t})dW_s + \int_0^t D_s[\sigma_s(A_{s,t})]ds, \quad 0 \le t \le 1.$$

Proof: Due to (2.2.3) and (2.2.6) we have

$$\sigma(A_{s,t})I_{[0,t]}(s) = \sum_{j=1}^n \frac{1}{\sqrt{t_j - t_{j-1}}} f_j(Y_{s,t}(\bar{\omega}))I_{[t_{j-1}\wedge t, t_j \wedge t]}(s), \ 0 \le s \le 1.$$

Clearly, we can apply Proposition 1.6 to the processes $(f_j(Y_{s,t}(\bar{\omega}))I_{[t_{j-1}\wedge t, t_j \wedge t]}(s)) \in \mathbb{L}_{C-}^{1,2}$. Taking into account that

$$(D_- f_j(Y_{s,t}(\bar{\omega})))_s (= \sum_{i=1}^n (\frac{\partial}{\partial x_i} f_j)(Y_{s,t}(\bar{\omega}))\frac{\partial}{\partial x_j} Y_{s,t}^i(\bar{\omega})) = D_s[\sigma_s(A_{s,t})], \ t_{j-1} \le s < t_j,$$

we obtain (2.2.13). An analogous discussion yields (2.2.12).

Now we can state:

Proposition 2.2.6 *For any* $(\sigma_s) \in \mathbb{L}^S$ *the densities*

$$\mathcal{L}_t = \frac{dP \circ [A_t]^{-1}}{dP}, \quad L_t = \frac{dP \circ [T_t]^{-1}}{dP} \quad and \quad L_{s,t} = L_s(A_{s,t})^{-1}L_t$$

are given by

(2.2.14)
$$\mathcal{L}_t = \exp\{-\int_0^t \sigma_s(T_s)dW_s - \frac{1}{2}\int_0^t \sigma_s(T_s)^2 ds$$

$$- \int_0^t \int_0^s (D_r\sigma_s)(T_s)D_s[\sigma_r(T_r)]drds\},$$

(2.2.15)
$$L_{s,t} = \exp\{\int_s^t \sigma_r(A_{r,t})dW_r - \frac{1}{2}\int_s^t \sigma_r(A_{r,t})^2 dr$$

$$- \int_s^t \int_r^t (D_u\sigma_r)(A_{r,t})D_r[\sigma_u(A_{u,t})]dudr\}$$

and

$$L_t = L_{o,t}, \quad for \ all \quad 0 \le s \le t \le 1.$$

Proof: For checking (2.2.14) it suffices to deduce from Proposition 2.1.5 that

$$D_s[\sigma_s(T_s\omega)] = (D_s\sigma_s)(T_s\omega) + \int\limits_0^s (D_r\sigma_s)(T_s\omega)D_s[\sigma_r(T_r\omega)]dr,$$

and to substitute this as well as the relation (2.2.12) in the formula for $\dfrac{dP \circ [A_t]^{-1}}{dP}$ given by Proposition 2.2.3, whereas (2.2.15) follows from the relations (2.2.11), (2.2.13) and Proposition 2.1.5,

$$D_s[\sigma_s(A_{s,t}\omega)] = (D_s\sigma_s)(A_{s,t}\omega) - \int\limits_s^t (D_r\sigma_s)(A_{s,t}\omega)D_s[\sigma_r(A_{r,t}\omega)]drds, \ 0 \le s \le t \le 1.$$

Remark: One can derive (2.2.14) also directly, without using Proposition 2.2.3. In this case Proposition 2.1.9, i.e., Kusuoka's theorem, has to be applied to the transformation

$$T_t\omega = \omega + \int\limits_0^{t\wedge\cdot} \sigma_r(T_r\omega)dr, \quad \omega \in \Omega.$$

It is not hard to show that the assumptions of Kusuoka's theorem are satisfied. So one easily concludes that the inverse transformation A_t of T_t has the density

$$\frac{dP \circ [A_t]^{-1}}{dP} = |d_c(-D_r[\sigma_s(T_s)])| \exp\{-\int\limits_0^1 \sigma_s(T_s)dW_s - \frac{1}{2}\int\limits_0^1 \sigma_s(T_s)^2 ds\}.$$

In [10], the Carleman-Fredholm determinant $|d_c(-D_r[\sigma_s(T_s)])|$ has been computed directly; here we can conclude from (2.2.14) that

$$|d_c(-D_r[\sigma_s(T_s)])| = \exp\{-\int\limits_0^1\int\limits_0^s (D_r\sigma_s)(T_s)D_s[\sigma_r(T_r)]drds\}.$$

Lemma 2.2.2 and Proposition 2.2.6 show that Theorem 2.2.1 holds for any process $(\sigma_s) \in \mathbb{L}^S$. Now we want to derive the statement of Theorem 2.2.1 for any process (σ_s) in $L^2([0,1], \mathbb{D}^{1,\infty})$. For this fix any such process (σ_s) and recall that, by virtue of Proposition 2.1.3, there exists a sequence $((\sigma_s^n))$ of $\mathbb{L}^S$ which approximates the process (σ_s) in $\mathbb{L}^{1,2}$ and is bounded in $L^2([0,1], \mathbb{D}^{1,\infty})$, say by a constant C_σ. Fix such a sequence $((\sigma_s^n))$. Due to Lemma 2.2.2 and Proposition 2.2.6 we know that Theorem 2.2.1 is true for any (σ_s^n). Denote by $\{T_t^n, 0 \le t \le 1\}$ the family of invertible absolutely continuous transformations associated with (σ_s^n) by (2.2.1), let $\{A_{s,t}^n, 0 \le s \le t \le 1\}$ be defined for (σ_s^n) by (2.2.4) and put $A_t^n = A_{0,t}^n$.
We prepare the proof of Theorem 2.2.1 by the following Proposition:

Proposition 2.2.7 *With the notations introduced above, the sequence of processes $((\sigma_s^n(T_s^n)))$ converges in $\mathbb{L}^{1,2}$ and is bounded in $L^2([0,1], \mathbb{D}^{1,\infty})$.*

For the proof of this proposition we need some auxiliary lemmata.

Lemma 2.2.8 *The sequence of processes $((\sigma_s^n(T_s^n)))$ is bounded in $L^2([0,1], \mathbb{D}^{1,\infty})$.*

Proof: Obviously, it holds

$$\int_0^1 \|\sigma_s^n(T_s^n)\|_\infty^2 ds \leq C_\sigma^2, \quad n = 1, 2, 3, \ldots$$

Thus, it remains to show that the sequence $(\int_0^1 \| \, |D[\sigma_t^n(T_t^n)]|_2\|_\infty dt)$ is bounded. Recall that $\sigma_t^n(T_t^n) \in \mathcal{S}$, $0 \leq t \leq 1$, i.e., we can apply Proposition 2.1.5 to $\sigma_t^n(T_t^n)$, and obtain

$$D_r[\sigma_t^n(T_t^n)] = (D_r\sigma_t^n)(T_t^n) + \int_0^t (D_s\sigma_t^n)(T_t^n) D_r[\sigma_s^n(T_s^n)] ds,$$

$$\text{for all} \quad 0 \leq r \leq 1.$$

Hence,

$$\int_0^t \| \, |D[\sigma_s^n(T_s^n)]|_2\|_\infty^2 ds \leq 2 \int_0^1 \| \, |D\sigma_s^n|_2\|_\infty^2 ds$$

$$+ 2 \int_0^t \| \, |D\sigma_s^n|_2\|_\infty^2 \int_0^s \| \, |D[\sigma_r^n(T_r^n)]|_2\|_\infty^2 dr ds, \ 0 \leq t \leq 1.$$

Thus, we can apply Gronwall's lemma and obtain

$$(2.2.16) \quad \int_0^1 \| \, |D[\sigma_s^n(T_s^n)]|_2\|_\infty^2 ds \leq 2 \int_0^1 \| \, |D\sigma_s^n|_2\|_\infty^2 ds \cdot \exp\{2 \int_0^1 \| \, |D\sigma_s|_2\|_\infty^2 ds\}$$

$$\leq 2C_\sigma^2 \exp\{2C_\sigma^2\}, \quad \text{for all} \quad n = 1, 2, 3, \ldots$$

Lemma 2.2.9 *The family $M = \{L_t^n = \dfrac{dP \circ [T_t^n]^{-1}}{dP}, 0 \leq t \leq 1, n = 1, 2, 3, \ldots\}$ of densities is uniformly integrable.*

Proof: For the uniform integrability of M if suffices to show that

$$(2.2.17) \qquad\qquad \sup_{0 \leq t \leq 1, n \geq 1} E[L_t^n | \ln L_t^n|] < \infty.$$

Set $\mathcal{L}_t^n = \dfrac{dP \circ [A_t^n]^{-1}}{dP}$. Then, $L_t^n = \mathcal{L}_t^n(A_t^n)^{-1}$, and

$$E[L_t^n | \ln L_t^n|] = E[L_t^n | \ln \mathcal{L}_t^n(A_t^n)|] = E[| \ln \mathcal{L}_t^n|].$$

Applying Proposition 2.2.6 we can estimate

$$(2.2.18) \qquad E[|\ln \mathcal{L}_t^n|] \;\leq\; E[|\int_0^t \sigma_s^n(T_s^n)dW_s|] + \frac{1}{2}E[\int_0^t \sigma_s^n(T_s^n)^2 ds]$$

$$+ E[|\int_0^t \int_0^s (D_r\sigma_s^n)(T_s^n)D_s[\sigma_r^n(T_r^n)]drds|]$$

Obviously,

$$\frac{1}{2}E[\int_0^t \sigma_s^n(T_s^n)^2 ds] \leq \frac{1}{2}C_\sigma^2, \quad 0 \leq t \leq 1, \; n = 1,2,3,\ldots.$$

Thus, it remains to estimate the first and the third term of the right-hand side of (2.2.18). For this, apply (2.2.16) to the first term,

$$E[|\int_0^t \sigma_s^n(T_s^n)dW_s|]^2 \;\leq\; E[\int_0^1 \sigma_s^n(T_s^n)^2 ds] + E(\int_0^1 \int_0^1 |D_r[\sigma_s^n(T_s^n)]|^2 drds]$$

$$\leq\; C_\sigma^2 + 2C_\sigma^2 \exp\{2C_\sigma^2\}, \quad 0 \leq t \leq 1, \; n = 1,2,3,\ldots,$$

and to the third term,

$$E[|\int_0^t \int_0^s (D_r\sigma_s^n)(T_s^n)D_s[\sigma_r^n(T_r^n)]drds|] \leq 2C_\sigma^2 \exp\{C_\sigma^2\}, \quad 0 \leq t \leq 1, \; n = 1,2,3,\ldots.$$

Since all these estimates are uniform relative to t and n, condition (2.2.17) is satisfied. This completes the proof.

Lemma 2.2.10 *The sequence $((\sigma_t^n(T_t^n)))$ converges in $L^2([0,1] \times \Omega)$.*

Proof: Applying Proposition 2.1.4 we obtain

$$E[\int_0^t |\sigma_s^n(T_s^n) - \sigma_s^m(T_s^m)|^2 ds] \leq 2E[\int_0^t |\sigma_s^n - \sigma_s^m|^2 L_s^n ds] +$$

$$+ 2E[\int_0^t \| |D\sigma_s^m|_2 \|_\infty^2 ds \cdot \int_0^s |\sigma_r^n(T_r^n) - \sigma_r^m(T_r^m)|^2 drds],$$

for $0 \leq t \leq 1$, $n,m = 1,2,3,\ldots$, where L_s^n denotes the density of the transformation T_s^n.

Consequently, Gronwall's lemma provides

$$E[\int_0^t |\sigma_s^n(T_s^n) - \sigma_s^m(T_s^m)|^2 ds]$$

$$\leq 2\exp\{2C_\sigma^2\}E[\int_0^1 |\sigma_s^n - \sigma_s^m|^2 L_s^n ds], \quad n,m = 1,2,3,\ldots.$$

Since $M = \{L_s^n, 0 \leq s \leq 1, n = 1, 2, 3, \ldots\}$ is uniformly integrable and $((\sigma_s^n))$ is convergent in $L^2([0,1] \times \Omega)$ and bounded in $L^2([0,1], L^\infty(\Omega))$, the right-hand side of this estimation converges to zero as $n \to \infty$. Consequently, we have the desired statement.

Lemma 2.2.10 allows us to introduce the process $(\bar{\sigma}_s)$ as the limit of the sequence $((\sigma_s^n(T_s^n)))$ in $L^2([0,1] \times \Omega)$. For each $0 \leq t \leq 1$ put

$$T_t\omega = \omega + \int_0^{t\wedge\cdot} \bar{\sigma}_s(\omega)ds, \qquad \omega \in \Omega.$$

Lemma 2.2.11 *For each $0 \leq t \leq 1$, the transformation T_t introduced above is absolutely continuous and*

$$\bar{\sigma}_t(\omega) = \sigma_t(T_t\omega), \qquad a.e.$$

Proof: The absolute continuity of the transformation T_t, $0 \leq t \leq 1$, follows immediately from Proposition 2.1.7 and the definition of $(\bar{\sigma}_s)$. Thus, by virtue of Proposition 2.1.8, we have

$$\bar{\sigma}_s = L^2(\Omega) - \lim_{n\to\infty} \sigma_t^n(T_t^n) = \sigma_t(T_t), \qquad a.e.$$

Now we are able to prove Proposition 2.2.7:

Proof of Proposition 2.2.7: Since the random variable $\sigma_t^n(T_t^n)$ belongs to $\mathcal{S}$ for any $0 \leq t \leq 1$, we can apply Proposition 2.1.5 to $\sigma_t^n(T_t^n)$. This yields

$$D_r[\sigma_t^n(T_t^n)] = (D_r\sigma_t^n)(T_t^n) + \int_0^t (D_s\sigma_t^n)(T_t^n)D_r[\sigma_s^n(T_s^n)]ds$$

for each $0 \leq r, t \leq 1$, $n = 1, 2, 3, \ldots$. Taking (2.2.16) into account we deduce

$$E[\int_0^1 |D[\sigma_t^n(T_t^n)] - D[\sigma_t^m(T_t^m)]|_2^2 dt]$$

$$\leq (2 + 8C_\sigma^2 \exp\{2C_\sigma^2\})\exp\{4C_\sigma^2\}E[\int_0^1 |(D\sigma_t^n)(T_t^n) - (D\sigma_t^m)(T_t^m)|_2^2 dt]$$

for all $n, m = 1, 2, 3, \ldots$. Hence, Proposition 2.1.8 together with the Lemmata 2.2.9 and 2.2.10 implies

$$\lim_{n,m\to\infty} E[\int_0^1 |D[\sigma_t^n(T_t^n) - D[\sigma_t^m(T_t^m)]|_2^2 dt] = 0.$$

Together with Lemma 2.2.10 this completes the proof.

Remark: By virtue of Lemma 2.2.11 it now becomes clear that $(\sigma_t(T_t))$ is the limit of the sequence $((\sigma_t^n(T_t^n)))$ in $\mathbb{L}^{1,2}$.

Finally we prove Theorem 2.2.1:

Proof of Theorem 2.2.1: We use the notations introduced above for the process $(\sigma_s) \in L^2([0,1], \mathbb{D}^{1,\infty})$ and its approximating sequence $((\sigma_s^n)) \subseteq \mathbb{L}^S$. From the Remark above we know that the family $\{T_t, 0 \le t \le 1\}$ of absolutely continuous transformations is a solution of equation (2.2.1). The uniqueness of such a family of absolutely continuous transformations follows immediately from Proposition 2.1.4.

Thus, it remains to check the invertibility of T_t and to compute the density of its inverse transformation A_t. For this return again to the approximating sequence $((\sigma_s^n)) \subset \mathbb{L}^S$, and denote by $\{A_{s,t}^n, 0 \le s \le t \le 1\}$ the family of transformations associated to (σ_s^n) by (2.2.4), i.e., for each $0 \le t \le 1$ and $\omega \in \Omega$ we have

$$(2.2.19) \qquad A_{s,t}^n \omega = \omega - \int_{s\wedge .}^{t\wedge .} \sigma_r^n(A_{r,t}^n \omega) dr, \quad 0 \le s \le t \le 1.$$

From the Remark to Proposition 2.2.3 and Proposition 2.2.6 we know that $A_{s,t}^n$ is absolutely continuous and invertible, and its inverse has the density

$$
\begin{aligned}
(2.2.20) \qquad L_{s,t}^n &= \exp\{\int_s^t \sigma_r^n(A_{r,t}^n) dW_r - \frac{1}{2} \int_s^t \sigma_r^n(A_{r,t}^n)^2 dr \\
&\quad - \int_s^t \int_r^t (D_u \sigma_r^n)(A_{r,t}^n) D_r[\sigma_u^n(A_{u,t}^n)] du\, dr\}.
\end{aligned}
$$

Obviously, the processes $(\sigma_s^n(A_{s,t}^n) I_{[0,t]}(s))$, $0 \le t \le 1$, are in $L^2([0,1], \mathbb{D}^{1,\infty})$, and from (2.2.19) we can derive in analogy to the proof of Lemma 2.2.8 that this family of sequences of processes is bounded there, uniformly with respect to $0 \le t \le 1$. Thus, in the same manner as in Lemma 2.2.9, we can show that the family of densities

$$\{\mathcal{L}_{s,t}^n = \frac{dP \circ [A_{s,t}^n]^{-1}}{dP}, \ 0 \le s \le t \le 1, n = 1, 2, 3, \ldots\}$$

verifies the relation

$$\sup_{s \le t, n \ge 1} E[\mathcal{L}_{s,t}^n | \ln \mathcal{L}_{s,t}^n |] = \sup_{s \le t, n \ge 1} E[|\ln L_{s,t}^n|] < \infty,$$

which implies its uniform integrability. Now, we only have to recall the arguments of the Lemmata 2.2.10 and 2.2.11 in order to see that, for each $0 \le t \le 1$, the sequence of processes $((\sigma_s^n(A_{s,t}^n) I_{[0,1]}(s)))$ converges in $L^2([0,1] \times \Omega)$ to some process $(\bar{\sigma}_s^t)$, and the transformation

$$(2.2.21) \qquad A_t \omega = \omega - \int_0^{t\wedge .} \bar{\sigma}_s^t(\omega) ds, \quad \omega \in \Omega,$$

is absolutely continuous. Applying Proposition 2.1.8 and the Remark to the proof of Proposition 2.2.7 we obtain

$$
\begin{aligned}
\bar{\sigma}_s^t &= L^2(\Omega) - \lim_{n\to\infty} \sigma_s^n(A_{s,t}^n) \\
&= L^2(\Omega) - \lim_{n\to\infty} [\sigma_s^n(T_s^n)] \circ A_t^n = \sigma_s(T_s A_t), \quad 0 \le s \le t \le 1,
\end{aligned}
$$

and

$$
\begin{aligned}
\bar{\sigma}_s^t(T_t) &= L^2(\Omega) - \lim_{n\to\infty} [\sigma_s^n(A_{s,t}^n)] \circ T_t^n \\
&= L^2(\Omega) - \lim_{n\to\infty} \sigma_s^n(T_s^n) = \sigma_s(T_s), \quad 0 \le s \le t \le 1.
\end{aligned}
$$

Taking into account (2.2.1) and (2.2.21), this implies for each $0 \le t \le 1$:

$$
A_t(T_t\omega) = T_t\omega - \int_0^{t\wedge\cdot} \bar{\sigma}_s^t(T_t\omega)ds = \omega, \quad \text{a.e.},
$$

and

$$
T_t(A_t\omega) = A_t\omega + \int_0^{t\wedge\cdot} \sigma_s(T_s A_t\omega)ds = \omega, \quad \text{a.e.},
$$

i.e., A_t is inverse to T_t.

Now, it only remains to prove formula (2.2.2). Recall from Proposition 2.2.6 that the transformation A_t^n has the density

$$
\begin{aligned}
\mathcal{L}_t^n &= \exp\{-\int_0^t \sigma_s^n(T_s^n)dW_s - \frac{1}{2}\int_0^t \sigma_s^n(T_s^n)^2 ds \\
&\quad - \int_0^t \int_0^s (D_r \sigma_s^n)(T_s^n) D_s[\sigma_r^n(T_r^n)]dr\, ds\},
\end{aligned}
$$

for each $n = 1, 2, 3, \ldots, 0 \le t \le 1$.

From Proposition 2.1.8 and the Remark to the proof of Proposition 2.2.7 we deduce that $(\mathcal{L}_t^n)$ converges in probability to the right-hand side of (2.2.2) as $n \to \infty$. Since, on the other hand, this sequence is uniformly integrable, this convergence is even in $L^1(\Omega)$. Note, by Proposition 2.1.8 we get for any $F \in \mathcal{S}$ that $F(A_t) = L^2(\Omega) - \lim_{n\to\infty} F(A_t^n)$. Thus, we have for each $0 \le t \le 1$,

$$
E[F(A_t)] = \lim_{n\to\infty} E[F(A_t^n)] = \lim_{n\to\infty} E[F\mathcal{L}_t^n] = E[F\mathcal{L}_t],
$$

i.e., (2.2.2) describes the density of A_t. This completes the proof.

Use the notations introduced above and set $A_{s,t} = T_s \circ A_t$, $0 \le s \le t \le 1$. From the proof of Theorem 2.2.1 and its arguments it is clear that Lemma 2.2.2 and Proposition 2.2.6 can be generalized to $(\sigma_s) \in L^2([0,1], \mathbb{D}^{1,\infty})$:

Proposition 2.2.12 *Let $(\sigma_s) \in L^2([0,1], I\!\!D^{1,\infty})$. Then $\{A_{s,t} = T_s \circ A_t,\ 0 \leq s \leq t \leq 1\}$ is the solution of the equation*

$$(2.2.22) \qquad A_{s,t}\omega = \omega - \int_{s\wedge\cdot}^{t\wedge\cdot} \sigma_r(A_{r,t}\omega)dr, \qquad a.e.,\ 0 \leq s \leq t,$$

which is unique in the class of families of absolutely continuous transformations. Moreover, the inverse transformation $A_{s,t}^{-1} = A_t \circ T_s$ has the density

$$(2.2.23) \qquad L_{s,t} = \exp\{\int_s^t \sigma_r(A_{r,t})dW_r - \frac{1}{2}\int_s^t \sigma_r(A_{r,t})^2 dr$$
$$- \int_s^t \int_r^t (D_u\sigma_r)(A_{r,t})D_r[\sigma_u(A_{u,t})]du\,dr\}, \qquad 0 \leq s \leq t \leq 1.$$

Finally we will present a lemma needed for the study of linear stochastic differential equations.

Lemma 2.2.13 *Let $F \in \mathcal{S}$ and $(\sigma_s) \in L^2([0,1], I\!\!D^{1,\infty})$. Denote by $\{T_t, 0 \leq t \leq 1\}$ the family of transformations associated to (σ_s) by (2.2.1) and by $\{A_{s,t}, 0 \leq s \leq t \leq 1\}$ the family determined by equation (2.2.22). Then it holds:*

(i) *The process $\{F(T_t), 0 \leq t \leq 1\}$ belongs to $L^2([0,1], I\!\!D^{1,\infty})$, is absolutely continuous with respect to the time parameter t, and*

$$\frac{d}{dt}F(T_t) = \sigma_t(T_t)(D_tF)(T_t), \qquad a.e.$$

(ii) *If, moreover, (σ_s) is assumed to belong to $I\!\!L^{\mathcal{S}}$, then, for any $0 \leq s \leq 1$, the process $(F(A_{s,t})I_{[s,t]}(t))$ is in $\widetilde{I\!\!L^{1,\infty}}$, and $F(A_{s,t}) \in \mathcal{S}$, $s \leq t \leq 1$. The function $t \mapsto F(A_{s,t}\omega)$ is pathwise differentiable, and*

$$(2.2.24) \qquad \frac{d}{dt}F(A_{s,t}) = -\sigma_t D_t[F(A_{s,t})], \qquad a.e.$$

Additionally, with the notations $B_F = \|DF\|_{L^\infty([0,1]\times\Omega)}$ and $B_\sigma = \|D\sigma\|_{L^\infty([0,1]^2\times\Omega)}$, we have

$$(2.2.25) \qquad \|D_t[F(A_{s,t})]\|_\infty \leq B_F(1 + B_\sigma\exp(B_\sigma)), \qquad 0 \leq s \leq t \leq 1.$$

Proof: The statement (i) can be checked easily by making use of the representation for smooth random variables $F \in \mathcal{S}$. Part (ii) follows from the representation of $F(A_{s,t})$ found in the proof of Lemma 2.2.4:

$$F(A_{s,t}\omega) = \varphi(Y_{s,t}(\bar\omega)), \quad \text{for some} \quad \varphi \in C_b^\infty(R^n),$$

where $Y_{s,t}(x)$ and $\bar{\omega}$ are defined by (2.2.6). In particular, the differentiability of $t \mapsto \varphi(Y_{s,t}(x))$ with derivative

$$\frac{d}{dt}[\varphi(Y_{s,t}(x))] = -f_t(x)\frac{d}{dx}[\varphi(Y_{s,t}(x))]$$

provides (2.2.24) if we substitute $x = \bar{\omega}$. Thus, it remains to check (2.2.25). For this note that, by virtue of Proposition 2.1.5 and equation (2.2.22),

$$(2.2.26) \qquad D_t[F(A_{s,t})] = (D_tF)(A_{s,t}) - \int_s^t (D_rF)(A_{s,t})D_t[\sigma_r(A_{r,t})]dr$$

and

$$D_t[\sigma_s(A_{s,t})] = (D_t\sigma_s)(A_{s,t}) - \int_s^t (D_r\sigma_s)(A_{s,t})D_t[\sigma_r(A_{r,t})]dr,$$

for each $0 \le s \le t \le 1$. Hence, Gronwall's lemma implies

$$|D_t[\sigma_s(A_{s,t})]| \le B_\sigma \exp(B_\sigma),$$

which, substituted in (2.2.26), provides the desired result.

2.3 Transformations given by a flow adapted to some enlarged filtration

The purpose of this section is to study the family $\{A_{s,t}, 0 \le s \le t \le 1\}$ of transformations of the Wiener space Ω into itself which satisfies the equation

$$(2.3.1) \qquad A_{s,t}\omega = \omega - \int_{s\wedge.}^{t\wedge.} \sigma_r(A_{r,t}\omega)dr, \quad \text{a.e., } 0 \le s \le t \le 1,$$

for a process (σ_s) for which the smoothness requirement of Section 2.2 is replaced by the assumption of adaptedness to a certain enlargement of the filtration $(\mathcal{F}_t^W)$ of the canonical process (W_t). This new assumption does not guarantee the invertibility of the transformations $A_{s,t}$, so that the approach will slightly differ from that of Section 2.2.

The main result of this section is the following:

Theorem 2.3.1 *Let* $G \in \widetilde{D^{1,\infty}}$, $a = \|G\|_\infty$ *and* $R_a = [-a, a]$. *Furthermore, let* $K.(.,.) : [0,1] \times \Omega \times R_a \to \mathbb{R}^1$ *be a measurable function satisfying the following assumptions:*

(i) $(K_s(x))$ *is* $(\mathcal{F}_s^W)$*-adapted for each* $x \in R^1$,

(ii) $K_s(\omega, .) \in C^2(R^1)$ *for a.e.* $(s, \omega) \in [0,1] \times \Omega$.

(iii) The norm

$$\||K\||_{(1)} \;=\; \|(\int_0^1 K_r(.,.)^2 dr)^{1/2}\|_{L^\infty(\Omega \times R_a)} + (\int_0^1 \|K_r'(.,.)\|_{L^\infty(\Omega \times R_a)}^2 dr)^{1/2}$$

$$+\|(\int_0^1 K_r''(.,.)^2 dr)^{1/2}\|_{L^4(\Omega \times R_a)}$$

is finite.

Then, there exists a unique family $\{A_{s,t}, 0 \le s \le t \le 1\}$ of absolutely continuous transformations $A_{s,t} : \Omega \to \Omega$ that satisfy the equation

$$(2.3.2) \qquad A_{s,t}\omega = \omega - \int_{s\wedge.}^{t\wedge.} K_r(\omega, G(A_{r,t}\omega))dr, \quad a.e., \quad 0 \le s \le t \le 1.$$

Moreover, the corrected integral of $(K_r(G(A_{r,t}))I_{[s,t]}(r))$ exists and, with the notation

$$(2.3.3) \qquad L_{s,t} \;=\; \exp\{\int_s^t K_r(G(A_{r,t}))\bar{d}W_r - \frac{1}{2}\int_s^t K_r(G(A_{r,t}))^2 dr$$

$$- \int_s^t K_r'(G(A_{r,t}))(D_r G)(A_{r,t})dr\},$$

it holds

$$E[F(A_{s,t})L_{s,t}] = E[F], \quad \text{for all} \quad F \in L^\infty(\Omega), \ 0 \le s \le t \le 1.$$

Remarks:

1. In the context of a process $(K_r(G))$ adapted to the natural filtration $(\mathcal{F}_t^W \vee \sigma\{G\})$ enlarged by some $G \in I\!\!D^{1,\infty}$ the corrected integral seems to be more adaquate than the Skorohod integral, whose domain is more concentrated on the processes which are either $(\mathcal{F}_t^W)$-adapted or in $I\!\!L^{1,2}$. However, we should not expect that the corrected integral $\int_s^t K_r(G(A_{r,t}))\bar{d}W_r$ can be considered as an Itô integral with respect to the enlarged filtration $(\mathcal{F}_t^W \vee \sigma\{G\})$: It may happen that $(K_r(G(A_{r,t})))$ is neither in $I\!\!L^{1,2}$ nor $(\mathcal{F}_t^W \vee \sigma(G))$-adapted.

2. If we denote $(K_s(G))$ by (σ_s), then equation (2.3.2) is nothing else but (2.3.1), and (2.3.3) turns out to be the pendant to (2.2.23), although $(K_r(G(A_{r,t}))I_{[s,t]}(r))$ is not necessarily in Dom δ. Note that, if $(K_r(\omega, x) = K_r(\omega))$ is an adapted process, then Theorem 2.3.1 reduces to the nonanticipative Girsanov theorem.

The proof of the theorem will be given in the form of a series of lemmata. First note that, if the process $(\sigma_s = K_s(G))$ belongs to $I\!\!L^S$, then the results of Section 2.2 become applicable. Thus, we can state:

Lemma 2.3.2 *If* $K.(.,.) : [0,1] \times \Omega \times R^1 \to R^1$ *has the form*

$$(2.3.4) \quad K_s(\omega, x) = \sum_{i=1}^{n} f_i(\omega_{t_1}, \ldots, \omega_{t_n}, x) I_{[t_{j-1}, t_i)}(s), \quad (s, \omega, x) \in [0,1] \times \Omega \times R^1,$$

for some partition $0 = t_o < \ldots < t_n = 1$, $(n \geq 1)$ *of the interval* $[0,1]$ *and satisfies (i), (ii) and (iii) of Theorem 2.3.1 for some* $f_1, \ldots, f_n \in C_b^\infty(R^{n+1})$, *and if* G *is a smooth random variable of* S, *then the statement of Theorem 2.3.1 is true.*

Proof: Note that $\sigma_r(\omega) = K_r(\omega, G(\omega))$, $(r, \omega) \in [0,1] \times \Omega$, defines a process of $\mathbb{L}^S$ such that we can make use of Lemma 2.2.2 and the Remark to Proposition 2.2.3. Hence, it only remains to substitute the following obvious relations in (2.2.11) for getting (2.3.3):

$$\sigma_r(A_{r,t}\omega) = K_r(\omega, G(A_{r,t}\omega)),$$
$$(D_r\sigma_r)(A_{r,t}\omega) = K_r'(\omega, G(A_{r,t}\omega))(D_r G)(A_{r,t}\omega).$$

In order to show that the general result of Theorem 2.3.1 holds, we have to proceed analogously as in Section 2.2 and to approximate the process $(\sigma_r = K_r(G))$ of Theorem 2.3.1 by a suitable sequence $((\sigma_r^n = K_r^n(G^n)))$ of $\mathbb{L}^S$. Note that, in analogy to the case of $(\sigma_r) \in L^2([0,1], \mathbb{D}^{1,\infty})$ one can deduce here:

Lemma 2.3.3 *If* $(K_s(\omega, x))$ *satisfies the assumptions (i), (ii) and (iii) of Theorem 2.3.1, then there exists a sequence* $((K_s^n(\omega, x)))$ *of processes of the form (2.3.4) which has the following properties:*

(i) There is a real $C > 0$ *such that*

$$\||K^n\||_{(1)} \leq C, \qquad n = 1, 2, 3, \ldots$$

(ii) The sequence $((K_s^n(\omega, x)))$ *converges to* $(K_s(\omega, x))$ *with respect to the norm*

$$\||K\||_{(2)} = \|(\int_0^1 K_s(.,0)^2 ds)^{1/2}\|_2 + \sum_{i=1}^{2} \|(\int_0^1 K_s^{(i)}(.,.)^2 ds)^{1/2}\|_{L^2(\Omega \times R_a)}.$$

Now let us fix any process $(K_s(\omega, x))$ *with the properties (i), (ii) and (iii) of Theorem 2.3.1 and also any sequence* $((K_s^n(\omega, x)))$ *associated to* $(K_s(\omega, x))$ *by Lemma 2.3.3. Moreover, fix any* $G \in \mathbb{D}^{1,\infty}$ *with* $\|G\|_\infty \leq a$, *and choose a sequence* $(G^n) \subset S$ *approximating* G *in* $\mathbb{D}^{1,2}$ *and bounded in* $\mathbb{D}^{1,\infty}$ *(cf. Proposition 2.1.2).*

For convenience we also introduce the following abbreviations:
We set

$$\sigma_r(\omega) = K_r(\omega, G(\omega)) \quad \text{and} \quad \sigma_r^n(\omega) = K_r^n(\omega, G^n(\omega)), \, n = 1, 2, 3, \ldots.$$

Let $A_{s,t}^n : \Omega \to \Omega$, $0 \leq s \leq t \leq 1$, be the transformations associated to (σ_r^n) by equation (2.3.1), put

$$X_{s,t}^n(\omega) = G^n(A_{s,t}^n\omega), \quad Y_{s,t}^n(\omega) = (D_s G^n)(A_{s,t}^n\omega),$$

and, finally, define

$$(2.3.5) \quad L^n_{s,t} = \exp\{ \int_s^t K^n_r(X^n_{r,t}) \bar{d}W_r - \frac{1}{2} \int_s^t K^n_r(X^n_{r,t})^2 dr - \int_s^t (K^n_r)'(X^n_{r,t}) Y^n_{r,t} dr \}.$$

Clearly,

$$E[F(A^n_{s,t}) L^n_{s,t}] = E[F], \text{ for all } F \in L^\infty(\Omega), \ 0 \le s \le t \le 1, \ n = 1, 2, 3, \ldots.$$

Proving the convergence of this approximation requires a series of steps, which is comparable with that for the proof of Theorem 2.2.1, although there are also great differences in the details caused by the possible non-Fréchet-differentiability of $(K_r(G))$ and the possible noninvertibility of the transformations $A_{s,t}$ we are looking for.

Lemma 2.3.4 *The set* $M = \{L^n_{s,t}, (L^n_{s,t})^{-1}, 0 \le s \le t \le 1, n = 1, 2, 3, \ldots\}$ *is bounded in* $L^p(\Omega)$, *for all real* $p > 1$.

Proof: By virtue of the form (2.3.5) of the densities $L^n_{s,t}$ and the existence of a uniform bound of the second and the third integral in their exponent, it suffices to show

$$\sup_{n \ge 1} E[\exp\{ \sup_{0 \le s \le t \le 1} p | \int_s^t K^n_r(X^n_{r,t}) \bar{d}W_r | \}] < \infty, \quad \text{for all } 1 < p < \infty.$$

By Lemma 2.2.4,

$$(2.3.6) \qquad\qquad X^n_{r,t} = G^n - \int_r^t K^n_u(X^n_{u,t}) Y^n_{u,t} du,$$

i.e., with the notations

$$I^n_{s,t}(x) = \int_s^t K^n_r(x) dW_r \quad \text{and} \quad (I^n_{s,t})'(x) = \int_s^t (K^n_r)'(x) dW_r, \ 0 \le s \le t \le 1, n = 1, 2, 3, \ldots$$

for the Itô integrals of $(K^n_r(x))$ and $((K^n_r)'(x))$, respectively, we obtain

$$(2.3.7) \qquad \int_s^t K^n_r(X^n_{r,t}) \bar{d}W_r = I^n_{s,t}(G) - \int_s^t (I^n_{s,u})'(X^n_{u,t}) \frac{d}{du} X^n_{u,t} du.$$

Since there is some real C not depending on n and t such that

$$\int_0^t |\frac{d}{du} X^n_{u,t}| du \le C, \quad \text{a.e.,} \quad 0 \le t \le 1, \ n = 1, 2, 3, \ldots,$$

it holds

$$\sup_{0 \le s \le t \le 1} | \int_s^t K^n_r(X^n_{r,t}) \bar{d}W_r | \le \sup_{\substack{0 \le s \le t \le 1, \\ |x| \le a}} |I^n_{s,t}(x)| + C \sup_{\substack{0 \le s \le t \le 1, \\ |x| \le a}} |(I^n_{s,t})'(x)|, \ n = 1, 2, 3, \ldots$$

so that the desired result is now an immediate consequence of the following auxiliary result.

Lemma 2.3.5 *Let $f : [0,1] \times \Omega \times R_a \to R^1$ be a measurable function such that*

(i) $(f_t(.,x))$ is $(\mathcal{F}_t^W)$-adapted for each $x \in R_a$ and belongs to $L^2([0,1] \times \Omega)$.

(ii) $f_t(\omega,.) \in C^1(R^1)$ a.e. and $(f_t'(\omega,x)) \in L^2([0,1] \times \Omega \times R_a)$.

Then, with the notations $I_{s,t}(x) = \int_s^t f_r(x)dW_r$, $I_{s,t}'(x) = \int_s^t f_r'(x)dW_r$, it holds

$$E[\exp\{\sup_{s \le t, |x| \le a} |I_{s,t}(x)|\}] \le AE[\exp\{B\int_0^1 f_r(0)^2 dr\}]^{1/2}+$$

$$+A(E[\int_{-a}^a \exp\{B\int_0^1 f_r(x)^2 dr\}dx])^{1/2} + AE[\int_{-a}^a (\int_0^1 |f_r'(x)|^2 dr)^2 dx]^{1/2},$$

where A and B are some reals that depend on a, but not on $(f_s(x,\omega))$.

Proof: For convenience, put $I_t(x) = I_{o,t}(x)$. Since

$$\exp\{\sup_{s \le t, |x| \le a} |I_{s,t}(x)|\} \le \sup_{0 \le t \le 1}(\exp\{2I_t(0)\} + \int_{-a}^a \exp\{4I_t(x)\}dx + 2\int_{-a}^a |I_t'(x)|^2 dx$$

$$+ \exp\{-2I_t(0)\} + \int_{-a}^a \exp\{-4I_t(x)\}dx),$$

the submartingale maximum estimation provides the asserted result.

According to Lemma 2.3.4 we can state:

Lemma 2.3.6 *For each $0 \le s \le t \le 1$, the sequence of random variables $(X_{s,t}^n)$ converges in $L^2(\Omega)$; the limit is denoted by $X_{s,t}$.*

Proof: Fix any small $\varepsilon > 0$, and then choose a sufficiently small $\delta > 0$ such that

$$(2.3.8) \qquad 4a^2 \sup_{s \ge t, n \ge 1} P\{L_{s,t}^n < \delta\} \le \varepsilon \quad \text{(cf. Lemma 2.3.4).}$$

Let $N_\varepsilon > 0$ be sufficiently large such that, for all $n \ge N_\varepsilon$,

$$(2.3.9) \qquad \frac{1}{\delta}E[|G^n - G|^2] \le \varepsilon, \qquad \text{and}$$

$$(2.3.10) \qquad \||K^n - K|\|_{(2)}^2 \le \varepsilon.$$

Taking into account that by (2.3.8) and (2.3.9)

$$\sup_{s \le t, n \ge 1} E[|G^n(A_{s,t}^n) - G(A_{s,t}^n)|^2] \le 2\varepsilon,$$

we deduce
$$(2.3.11) \qquad E[|X_{s,t}^n - X_{s,t}^m|^2] \le 12\varepsilon + E[|G(A_{s,t}^n) - G(A_{s,t}^m)|^2].$$

Hence, by using Proposition 2.1.4, an estimation of the right-hand side of (2.3.11) yields

(2.3.12)

$$E[|X^n_{s,t} - X^m_{s,t}|^2] \leq 12\varepsilon + \|(\int_0^1 |D_r G|^2 dr)^{1/2}\|_\infty \cdot E[\int_s^t |K^n_r(X^n_{r,t}) - K^m_r(X^m_{r,t})|^2 dr] \leq$$

$$\leq 12\varepsilon(1 + \|(\int_0^1 |D_r G|^2 dr)^{1/2}\|_\infty)$$

$$+ 3\|(\int_0^1 |D_r G|^2 dr)^{1/2}\|_\infty \cdot E[\int_s^t |K_r(X^n_{r,t}) - K_r(X^m_{r,t})|^2 dr].$$

Here we have used (2.3.10) in order to obtain the latter line.
Now Gronwall's lemma allows us to deduce that for some real $C > 0$,

$$E[|X^n_{s,t} - X^m_{s,t}|^2] \leq 12\varepsilon(1 + \|(\int_0^1 |D_r G|^2 dr)^{1/2}\|_\infty) \exp(3C\|(\int_0^1 |D_r G|^2 dr)^{1/2}\|_\infty),$$
$$n, m \geq N_\varepsilon,$$

i.e., $(X^n_{s,t})$ forms a Cauchy-sequence in $L^2(\Omega)$ for all $0 \leq s \leq t \leq 1$.

Remark: Lemma 2.3.6 and the properties of the approximating sequence $((K^n_s(x, \omega)))$ make clear that the sequence of processes $(K^n_s(X^n_{s,t})I_{[0,t]}(s))$ tends to $(K_s(X_{s,t})I_{[0,t]}(s))$ in $L^2([0,1] \times \Omega)$ for any $0 \leq t \leq 1$.

Lemma 2.3.7 *For each $0 \leq t \leq 1$, the sequence of processes $(Y^n_{s,t}I_{[0,t]}(s))$ has some limit $(Y_{s,t}I_{[0,t]}(s))$ in $L^2([0,1] \times \Omega)$.*

Proof: Fix any small $\varepsilon > 0$ and any $\delta = \delta(\varepsilon) > 0$ satisfying (2.3.8). Then, choose a natural $N_\varepsilon > 0$ such that, for all $m, n \geq N_\varepsilon$,

$$\frac{1}{\delta} E[\int_0^1 |D_s G^n - D_s G^m|^2 ds] \leq \varepsilon,$$

and

$$E[\int_0^t |K^n_r(X^n_{r,t}) - K^m_r(X^m_{r,t})|^2 dr] \leq \varepsilon.$$

Then standard arguments already used in the proof of the preceding lemma show that, for any $n, m, k \geq N_\varepsilon$, it holds

$$E[\int_0^t |Y^n_{s,t} - Y^m_{s,t}|^2 ds] \leq 8\varepsilon + 3E[\int_0^t |(D_r G^k)(A^n_{r,t}) - (D_r G^k)(A^m_{r,t})|^2 ds]$$

$$\leq \quad 8\varepsilon + 3\int_0^t \| \int_0^1 |D_s D_r G^k|^2 ds \|_\infty \cdot E[|K_r^n(X_{r,t}^n) - K_r^m(X_{r,t}^m)|^2] dr$$

$$\leq \quad 8\varepsilon + 3\| \int_0^1 |D_s D_\bullet G^k|^2 ds \|_{L^\infty([0,1]\times\Omega)} \times$$

$$\times E[\int_0^t |K_r^n(X_{r,t}^n) - K_r^m(X_{r,t}^m)|^2 dr]$$

$$\leq \quad (8 + 3\| \int_0^1 |D_s D_\bullet G^k|^2 ds \|_{L^\infty([0,1]\times\Omega)}) \cdot \varepsilon.$$

The latter expression is small if ε is small, i.e., if only n and m are large enough. This completes the proof.

The Lemmata 2.3.6 and 2.3.7 and the Remark to Lemma 2.3.6 allow to pass to infinity in equation (2.3.6), and so we see that the process $(X_{s,t})$ is absolutely continuous with respect to s for each fixed $0 \leq t \leq 1$, and

$$\frac{d}{ds} X_{s,t} \quad = \quad Y_{s,t} \cdot K_s(X_{s,t}), \qquad 0 \leq s \leq t,$$
$$X_{t,t} \quad = \quad G.$$

Remark that the two processes $(X_{s,t})$ and $(\frac{d}{ds} X_{s,t})$ are essentially bounded, namely,

$$\|X_{s,t}\|_\infty \leq a, \qquad 0 \leq s \leq t \leq 1,$$

and

$$\int_0^t \| \frac{d}{ds} X_{s,t} \|_\infty^2 ds \leq \sup_n \int_0^1 \| D_r G^n \|_\infty^2 dr.$$

This follows from the corresponding properties of the sequences $((X_{s,t}^n))$ and $((\frac{d}{ds} X_{s,t}^n))$. Consequently, the properties of $(K_s(\omega, x))$ required in Theorem 2.3.1 guarantee the existence of the corrected integral of $(K_r(X_{r,t}) I_{[s,t]}(r))$, for all $0 \leq s \leq t \leq 1$ and, moreover, with the notation $I_{s,t}(x) = \int_s^t K_r(x) dW_r$, the corrected integral satisfies the following relation

$$(2.3.13) \qquad \int_s^t K_r(X_{r,t}) \bar{d}W_r = I_{s,t}(G) - \int_s^t I'_{s,u}(X_{u,t}) \frac{d}{du} X_{u,t} du.$$

This equality allows us to show the following:

Lemma 2.3.8 *For all $0 \leq s \leq t \leq 1$ it holds*

$$\int_s^t K_r(X_{r,t}) \bar{d}W_r = L^1(\Omega) - \lim_{n\to\infty} \int_s^t K_r^n(X_{r,t}^n) \bar{d}W_r.$$

Proof: By virtue of the Remark above it suffices to prove that for all $0 \leq s \leq t \leq 1$,

$$I_{s,t}(G) = L^1(\Omega) - \lim_{n\to\infty} I^n_{s,t}(G^n),$$

and

$$(I'_{s,u}(X_{u,t})I_{[s,t]}(u)) = L^2([0,1] \times \Omega) - \lim_{n\to\infty}((I^n_{s,u})'(X^n_{u,t})I_{[s,t]}(u)).$$

We only prove of the second convergence, the proof of the first one runs analogously. Now, for the proof of the second statement, note that

(2.3.14)

$$E[\int_s^t |I'_{s,u}(X_{u,t}) - (I^n_{s,u})'(X^n_{u,t})|^2 du] \leq$$

$$\leq 4\{E[\int_s^t |K'_r(0) - (K^n_r)'(0)|^2 dr] + E[\int_s^t \int_{-a}^a |K'_r(x) - (K^n_r)'(x)|^2 dx dr]$$

$$+ E[\int_s^t \int_{-a}^a |K''_r(x) - (K^n_r)''(x)|^2 dx dr]\} + 2E[\int_s^t |I'_{s,u}(X_{u,t}) - I'_{s,u}(X^n_{u,t})|^2 du].$$

Here the latter expression tends to zero, since $I'_{s,u}(.)$ has a version in $C(R^1)$ and $\sup_{s\leq u, |x|\leq a} |I'_{s,u}(x)| \in L^2(\Omega)$ (cf. Lemma 2.3.5); the convergence to zero of the other terms of the right-hand side of (2.3.14) is guaranteed by the definition of the sequence $((K^n_s(\omega, x)))$.

By the Lemmata 2.3.4 - 2.3.8 we may conclude that, for all $0 \leq s \leq t \leq 1$ and for all $1 < p < \infty$, the sequence $(L^n_{s,t})$ of densities converges in $L^p(\Omega)$ to

$$(2.3.15) \quad L_{s,t} = \exp\{\int_s^t K_r(X_{r,t})\bar{d}W_r - \frac{1}{2}\int_0^t K_r(X_{r,t})^2 dr - \int_s^t K'_r(X_{r,t})Y_{r,t} dr\}.$$

Denote by $A_{s,t} : \Omega \to \Omega$ the transformation

$$(2.3.16) \qquad A_{s,t}\omega = \omega - \int_{s\wedge.}^{t\wedge.} K_r(\omega, X_{r,t}(\omega)) dr, \quad \omega \in \Omega, \; 0 \leq s \leq t \leq 1,$$

and recall that, in virtue of the Remark to Lemma 2.3.6, we have

$$F(A_{s,t}) = \lim_{n\to\infty} F(A^n_{s,t}); \quad \text{a.e. for all} \quad F \in \mathcal{S}.$$

Consequently,

$$E[F(A_{s,t})L_{s,t}] = \lim_{n\to\infty} E[F(A^n_{s,t})L^n_{s,t}] = E[F], \text{ for all } F \in \mathcal{S}, \; 0 \leq s \leq t \leq 1.$$

In particular, this implies the equivalence of $P \circ [A_{s,t}]^{-1}$ to the Wiener measure P.

Recall, the proper reason of this section is to prove Theorem 2.3.1. Well, this has almost been done already. It only remains to show that $L_{s,t}$ defined by (2.3.15) has the form (2.3.3) and that the family of absolutely continuous transformations $A_{s,t}$ defined by (2.3.16) is the unique solution of (2.3.2). For this note that, taking into account the Remark to Proposition 2.1.8 and Lemma 2.3.4, it follows

$$G(A_{s,t}) = L^2(\Omega) - \lim_{n \to \infty} G^n(A_{s,t}^n) = L^2(\Omega) - \lim_{n \to \infty} X_{s,t}^n = X_{s,t},$$

and

$$\begin{aligned}
((D_s G)(A_{s,t})) &= L^2([0,1] \times \Omega) - \lim_{n \to \infty} (D_s G^n)(A_{s,t}^n) \\
&= L^2([0,1] \times \Omega) - \lim_{n \to \infty} (Y_{s,t}^n) = (Y_{s,t}).
\end{aligned}$$

Hence, (2.3.15) and (2.3.3) coincide, and $\{A_{s,t}, 0 \le s \le t \le 1\}$ is a solution of equation (2.3.2). The uniqueness can be proved by applying Proposition 2.1.4. Hence, Theorem 2.3.1 has been proved.

2.4 Transformations defined by a smooth shift

In this section we study the transformation

$$(2.4.1) \qquad T\omega = \omega + \int_0^{\bullet} K_s(\omega)ds, \qquad \omega \in \Omega,$$

of Ω into itself for shift processes $(K_s) \in \mathbb{L}^{1,2}$ which are possibly anticipating. The main result of this section is the following:

Theorem 2.4.1 *Let $(K_s) \in \mathbb{L}^{1,2}$. Then, the transformation*

$$T\omega = \omega + \int_0^{\bullet} K_s(\omega)ds$$

is absolutely continuous if

(i) $\|(\int_0^1 \int_0^1 |D_t K_s|^2 ds\,dt)^{1/2}\|_\infty < 1,$
or

(ii) there is a $q > 1$ with $E[\exp\{\frac{q}{2} \int_0^1 K_s^2 ds\}] < \infty$ and $(\int_0^1 \int_0^1 |D_t K_s|^2 ds\,dt)^{1/2} < 1$ a.e. such that

$$(2.4.2) \qquad \mathcal{E} = \frac{(\int_0^1 \int_0^s |D_u K_s|^2 du\,ds)^{1/2} \cdot (\int_0^1 \int_s^1 |D_u K_s|^2 du\,ds)^{1/2}}{1 - (\int_0^1 \int_0^1 |D_u K_s|^2 du\,ds)^{1/2}} \in L^1(\Omega).$$

If both, (i) and (ii), are satisfied, then T is invertible, its inverse transformation A has the density

$$\mathcal{L} = |d_c(-DK)| \exp\{-\int_0^1 K_s dW_s - \frac{1}{2}\int_0^1 K_s^2 ds\},$$

with the Carleman-Fredholm determinant

(2.4.3) $$|d_c(-DK)| = \exp\{-\int_0^1 \int_0^s D_s K_r D_r[K_s(A_s)](T_s) dr\, ds\},$$

where $T_t\omega = \omega + \int_0^{t\wedge\cdot} K_s(\omega)ds$ and $A_t = T_t^{-1}$.

Remarks:

1. If (K_s) is adapted either to the filtration $(\mathcal{F}_t^W)$ generated by the canonical process (W_t) or to the filtration of the backward Wiener process $(W_t - W_1)$, then $\mathcal{E}$ vanishes.

2. For the Carleman-Fredholm determinant there is also another representation which is equivalent to (2.4.3):

    (2.4.4) $$|d_c(-DK)| = \exp\{-\int_0^1 \int_0^s D_r K_s D_s[K_r(A_s)](T_s) dr\, ds\}.$$

    We will return to this equivalence in connection with the proof of Theorem 2.4.1.

For the derivation of Theorem 2.4.1 we will use the results of Section 2.2, and therefore the transformation T will be embedded into a flow of transformations

(2.4.5) $$T_t\omega = \omega + \int_0^{t\wedge\cdot} \sigma_s(T_s\omega)ds, \quad \omega \in \Omega,\ 0 \le t \le 1,$$

such that, for some $(\sigma_s) \in L^2([0,1], \mathbb{D}^{1,\infty})$,

$$T_1\omega = T\omega.$$

The price for this approach to the proof of Theorem 2.3.1 is the requirement $(K_s) \in L^2([0,1], \mathbb{D}^{1,\infty})$; the general case is studied in [10].
By virtue of Theorem 2.2.1 we know the following:

Lemma 2.4.2 *Let $(K_t) \in L^2([0,1], \mathbb{D}^{1,\infty})$ and suppose that there exists a $(\sigma_s) \in L^2([0,1], \mathbb{D}^{1,\infty})$ such that the transformation T_1 defined by the stochastic flow (2.4.5) coincides with*

$$T\omega = \omega + \int_0^{\bullet} K_s(\omega)ds, \quad \omega \in \Omega.$$

Then T is absolutely continuous and invertible, and its inverse transformation A has the density given in Theorem 2.4.1.

Thus, we have to construct the process (σ_s). For this consider any arbitrary but fixed $(K_s) \in I\!\!L^{1,2}$ with $\|(\int_0^1 \int_0^1 |D_s K_t|^2 ds\, dt)^{1/2}\|_\infty < 1$. Proposition 2.1.3 allows us to choose any sequence $((K_s^n))$ of smooth step processes that converges to (K_s) in $I\!\!L^{1,2}$ and satisfies the following boundedness conditions (B):

(B.1) $(\int_0^1 \|K_s^n\|_\infty^2 ds)^{1/2} \leq (\int_0^1 \|K_s\|_\infty^2 ds)^{1/2}, \; n = 1, 2, 3, \ldots,$

(B.2) $a_K = \sup_{n \geq 1}(\int_0^1 \|(\int_0^1 |D_r K_s^n|^2 dr)^{1/2}\|_\infty ds)^{1/2} < \infty$

   and

(B.3) $c_K = \sup_{n \geq 1} \|(\int_0^1 \int_0^1 |D_r K_s^n|^2 dr\, ds)^{1/2}\|_\infty < 1.$

Fix any natural n and any $0 \leq t \leq 1$, and regard the transformation

$$T_t \omega = \omega + \int_0^{t \wedge \cdot} K_s^n(\omega) ds, \qquad \omega \in \Omega.$$

Since (K_s^n) belongs to $I\!\!L^S$ and has the property (B.3), we can apply the Fixed Point Theorem in order to deduce the invertibility of $T_t^n : \Omega \to \Omega$. Denote the inverse transformation by A_t^n. As in Section 2.2, here too, it is easy to see that

$$\sigma_t^n(\omega) = K_t^n(A_t^n \omega)$$

belongs to S for each $0 \leq t \leq 1$. By virtue of

(2.4.6) $(\int_0^1 \|\sigma_t^n\|_\infty^2 dt)^{1/2} \leq (\int_0^1 \|K_t\|_\infty^2 dt)^{1/2},$

the process (σ_t^n) belongs to $L^2([0,1], L^\infty(\Omega))$. For estimating the Fréchet derivative of (σ_t^n) we take into account that the transformation A_t^n satisfies the equation

$$A_t^n \omega = \omega - \int_0^{t \wedge \cdot} K_r^n(A_t^n \omega) dr, \qquad \omega \in \Omega,$$

and make use of Proposition 2.1.5:

(2.4.7) $D_r[K_s^n(A_t^n)] = (D_r K_s^n)(A_t^n) - \int_0^t (D_u K_s^n)(A_t^n) D_r[K_u^n(A_t^n)] du$

                        for all $0 \leq r, s, t \leq 1,$

i.e.,

$$\left(\int_0^1 \int_0^t |D_r[K_s^n(A_t^n)]|^2 dr\, ds\right)^{1/2} \le \frac{c_K}{1 - c_K}.$$

Since (2.4.7) holds for all $0 \le r, s, t \le 1$, we can put $s = t$. This provides

$$(2.4.8) \qquad \left(\int_0^1 \|\int_0^1 |D_r[\sigma_t^n]|^2 dr\|_\infty dt\right)^{1/2} \le a_K + \frac{a_K c_K}{1 - c_K}.$$

Consequently, $(\sigma_s^n) \in L^2([0,1], \mathbb{D}^{1,\infty})$. In particular, this guarantees that the transformations T_t^n and A_t^n are absolutely continuous (cf. Theorem 2.2.1). Further, note that the sequence of processes (σ_s^n) is bounded in $L^2([0,1], \mathbb{D}^{1,\infty})$. Recall that just this argument was used to deduce Lemma 2.2.9. Therefore, also the family of densities

$$(2.4.9) \qquad \{L_t^n = \frac{dP \circ [T_t^n]^{-1}}{dP} : 0 \le t \le 1,\ n = 1, 2, 3, \ldots\}$$

is uniformly integrable. On the other hand, the boundedness of $((\sigma_s^n))$ in $L^2([0,1], \mathbb{D}^{1,\infty})$ also implies the uniform integrability of the set

$$(2.4.10) \qquad \{\mathcal{L}_t^n = \frac{dP \circ [A_t^n]^{-1}}{dP} : 0 \le t \le 1,\ n = 1, 2, 3, \ldots, \}.$$

This allows us to establish the following:

Lemma 2.4.3 *Under the above assumptions the transformation T_t is absolutely continuous and invertible, its inverse transformation is absolutely continuous.*

Proof: Since the family $\{L_t^n = \frac{dP \circ [T_t^n]^{-1}}{dP}, \ n = 1, 2, 3, \ldots\}$ is uniformly integrable and (K_s) is the $\mathbb{L}^{1,2}$-limit of $((K_s^n))$, we can apply Proposition 2.1.7 in order to derive that, for any $0 \le t \le 1$, the transformation

$$T_t \omega = \omega + \int_0^{t \wedge \cdot} K_s(\omega) ds, \quad \omega \in \Omega,$$

is absolutely continuous. Let us now check whether $T_t : \Omega \to \Omega$ is invertible. Choose first any small $\varepsilon > 0$ and a real $M_\varepsilon > 0$ such that

$$2\left(\int_0^1 \|K_s\|_\infty^2 ds\right)^{1/2} \sup_{0 \le t \le 1, n \ge 1} \left(\int_{\{\mathcal{L}_t^n \ge M_\varepsilon\}} \mathcal{L}_t^n dP\right)^{1/2} \le \varepsilon \quad \text{(cf. (2.4.10))}.$$

Then, Proposition 2.1.4 provides the following estimates,

$$(E[|A_t^n - A_t^m|_H^2])^{1/2} \le$$

$$\le (E[\int_0^t |K_s^n(A_t^n) - K_s^n(A_t^m)|^2 ds])^{1/2} + (E[\int_0^t |K_s^n(A_t^m) - K_s^m(A_t^m)|^2 ds])^{1/2}$$

$$\le c_K (E[|A_t^n - A_t^m|_H^2])^{1/2} + (\varepsilon + \sqrt{M_\varepsilon}(E[\int_0^1 |K_s^n - K_s^m|^2 ds])^{1/2},$$

which show that

$$E[|A_t^n - A_t^m|_H^2] \to 0, \quad \text{as} \quad n, m \to \infty,$$

or, equivalently, that there exists a process $(K_s^t I_{[0,t]}(s)) \in L^2([0,1] \times \Omega)$ with

(2.4.11) $$E[\int_0^t |K_s^n(A_t^n) - K_s^t|^2 ds] \to 0, \quad \text{as} \quad n \to \infty.$$

Now we define the transformation

$$A_t \omega = \omega - \int_0^{t \wedge \cdot} K_s^t(\omega) ds, \qquad \omega \in \Omega,$$

and note that by Proposition 2.1.7 this transformation is absolutely continuous, and by Proposition 2.1.8

(2.4.12) $$K_s^t(\omega) = K_s(A_t \omega), \quad \text{a.e.,} \quad 0 \le s \le t \le 1.$$

Consequently,

(2.4.13) $$A_t \omega = \omega - \int_0^{t \wedge \cdot} K_s(A_t \omega) ds, \quad \text{a.e.}$$

In order to complete the proof we only have to show that the inverse of A_t exists and coincides with T_t. So we apply Proposition 2.1.8 once again, and obtain for a.e. $0 \le s \le t$

$$K_s^t(T_t \omega) = L^2(\Omega) - \lim_{n \to \infty} (K_s^n \circ A_t^n)(T_t^n \omega) = L^2(\Omega) - \lim_{n \to \infty} K_s^n(\omega) = K_s(\omega),$$

i.e.,

$$A_t \circ T_t \omega = T_t \omega - \int_0^{t \wedge \cdot} K_s^t(T_t \omega) ds = \omega, \quad \text{a.e.,} \quad 0 \le t \le 1.$$

On the other hand, in virtue of (2.4.13) we also obtain

$$T_t \circ A_t \omega = A_t \omega + \int_0^{t \wedge \cdot} K_s(A_t \omega) ds = \omega, \quad \text{a.e.,} \quad 0 \le t \le 1.$$

Consequently, A_t is the inverse transformation to T_t. This completes the proof.

For the transformation A_t introduced above we can state the following:

Lemma 2.4.4 *For each* $0 \le t \le 1$ *the process* $(K_s(A_t) I_{[0,t]}(s))$ *belongs to* $L^2([0,1], \mathbb{D}^{1,\infty})$.

Proof: Fix any $0 \le t \le 1$. In virtue of (B.3), by means of Proposition 2.1.8 we may deduce from equation (2.4.7) that the sequence $((D_r[K_s^n(A_t^n)] I_{[0,t]}(s)))$ converges in $L^2([0,1]^2 \times \Omega)$. On the other hand, since the sequence $((K_s^n(A_t^n) I_{[0,t]}(s)))$ is bounded in $L^2([0,1], I\!\!D^{1,\infty})$ according to (B.1), (B.2) and (2.4.7), its limit $(K_s^t = K_s(A_t)I_{[0,t]}(s))$ (cf. (2.4.11) and (2.4.12)) belongs to $L^2([0,1], I\!\!D^{1,\infty})$, too.

Now put $\sigma_t(\omega) = K_t(A_t\omega)$, $\omega \in \Omega$, $0 \le t \le 1$. Clearly, taking into account the Lemmata 2.4.3 and 2.4.4, we can derive from Proposition 2.1.5 that the random variable σ_t is in $I\!\!D^{1,\infty}$ for a.e. $0 \le t \le 1$,. Thus, a repetition of the estimates (2.4.6) - (2.4.8) now for $(K_s(A_t)I_{[0,t]}(s))$ and (σ_t), respectively, shows that

$$(2.4.14) \qquad\qquad (\sigma_t = K_t(A_t)) \in L^2([0,1], I\!\!D^{1,\infty}).$$

Note that $\sigma_t(T_t\omega) = K_t(\omega)$, a.e., $0 \le t \le 1$. However, this means that we now have the classical situation of a process $(\sigma_t) \in L^2([0,1], I\!\!D^{1,\infty})$ which the flow of transformations

$$T_t\omega = \omega + \int_0^{t\wedge\cdot} \sigma_s(T_s\omega)ds, \quad \text{a.e.}, \quad 0 \le t \le 1,$$

is associated to, where $T_1\omega = T\omega$, a.e. Hence, we can apply Theorem 2.2.1, which yields the following statement:

Proposition 2.4.5 *Let* $(K_t) \in L^2([0,1], I\!\!D^{1,\infty})$ *such that* $\| \int_0^1 \int_0^1 |D_s K_t|^2 ds\, dr \|_\infty < 1$. *Then the transformation* $T : \Omega \to \Omega$,

$$T\omega = \omega + \int_0^{\cdot} K_s(\omega)ds, \quad \omega \in \Omega,$$

is absolutely continuous and invertible, its inverse transformation has the density $\mathcal{L}$ *presented in Theorem 2.4.1.*

Remark 1 Note that the above derivation provides the Carleman-Fredholm determinant of the form (2.4.3) for $\mathcal{L}$. This Carleman-Fredholm determinant coincides with the expression (2.4.4). In order to check it, consider first a smooth step process (K_s^n) of the approximating sequence $((K_s^n))$ for (K_s). For this, let us use the notations introduced above. Then Proposition 2.1.5 yields

$$D_s[K_s^n(A_s^n)] - (D_s K_s^n)(A_s^n) = -\int_0^s (D_r K_s^n)(A_s^n)D_s[K_r^n(A_s^n)]dr,$$

i.e.,

$$(2.4.15) \quad D_s[K_s^n(A_s^n)](T_s^n) - D_s K_s^n = -\int_0^s D_r K_s^n \cdot D_s[K_r^n(A_s^n)](T_s^n)dr, \quad 0 \le s \le 1.$$

On the other hand,

$$(2.4.16) \qquad D_s[K_s^n(A_s^n)](T_s^n) - D_s K_s^n = (D_s\sigma_s^n)(T_s^n) - D_s[\sigma_s^n(T_s^n)],$$

and since $(\sigma_s^n = K_s^n(A_s^n)) \in \mathcal{S}$, $0 \le s \le 1$, we can apply Proposition 2.1.5 again. So we obtain

$$(2.4.17) \quad (D_s \sigma_s^n)(T_s^n) - D_s[\sigma_s^n(T_s^n)] = -\int_0^s (D_r \sigma_s^n)(T_s^n) D_s[\sigma_r^n(T_r^n)]dr, \quad 0 \le s \le 1.$$

Due to (2.4.16) the right-hand sides of (2.4.15) and (2.4.17) coincide, i.e.,

$$-\int_0^1 \int_0^s D_r K_s^n \cdot D_s[K_r^n(A_r^n)](T_s^n)dr\, ds \;=\; -\int_0^1 \int_0^s D_s[\sigma_r^n(T_r^n)](D_r \sigma_s^n)(T_s^n)dr\, ds$$

$$=\; -\int_0^1 \int_0^s D_s K_r^n \cdot D_r[K_s^n(A_s^n)](T_s^n)dr\, ds.$$

Using the convergence of $((K_s^n))$ and $((K_s^n(A_s^n)))$ in $\mathbb{L}^{1,2}$ to (K_s) and $(K_s(A_s))$, respectively, we can pass to the limit at the right-hand side, whereas the passage to the limit at the left-hand side is explained in [10]. Thus, we obtain

$$(\ln|d_c(-DK)| \;=\;) \; -\int_0^1 \int_0^s D_r K_s \cdot D_s[K_r(A_s)](T_s)dr\, ds$$

$$=\; -\int_0^1 \int_0^s D_s K_r \cdot D_r[K_s(A_s)](T_s)dr\, ds, \quad \text{a.e.}$$

Remark 2 In [10] Proposition 2.4.5 is proved for processes $(K_s) \in \mathbb{L}^{1,2}$ that have only been imposed to satisfy the assumptions (i) and (ii) of Theorem 2.4.1, and for the absolute continuity of T only condition (i) is needed (cf. Proposition 4.1 and Theorem 4.9 [10]).

Thus, for the proof of Theorem 2.4.1 it remains to show:

Proposition 2.4.6 *Let $(K_s) \in \mathbb{L}^{1,2}$ be chosen such that, for some real $q > 1$,*

$$(2.4.18) \qquad\qquad E[\exp\{\frac{q}{2}\int_0^1 K_s^2 ds\}] < \infty,$$

and $(D_r K_s)$ satisfies

$$(2.4.19) \qquad \int_0^1 \int_0^1 |D_r K_s|^2 dr\, ds < 1, \quad \text{a.e., as well as}$$

$$\mathcal{E} = \frac{(\int_0^1 \int_0^s |D_r K_s|^2 dr\, ds)^{1/2}(\int_0^1 \int_s^1 |D_r K_s|^2 dr\, ds)^{1/2}}{1 - (\int_0^1 \int_0^1 |D_r K_s|^2 dr\, ds)^{1/2}} \in L^1(\Omega).$$

Then the transformation $T : \Omega \to \Omega$,

$$T\omega = \omega + \int_0^{\cdot} K_s(\omega)ds, \quad \omega \in \Omega,$$

is absolutely continuous.

For the proof we need some results of [10], which will be presented in the following two statements.

Proposition 2.4.7 *Let* $(K_s) \in \mathbb{L}^{1,2}$ *be chosen such that*

$$c_K = \|(\int_0^1 \int_0^1 |D_r K_s|^2 dr \, ds)^{1/2}\|_\infty < 1.$$

Then, for any $0 \le t \le 1$, *the transformation*

$$T_t\omega = \omega + \int_0^{t \wedge \cdot} K_s(\omega)ds, \quad \omega \in \Omega,$$

is absolutely continuous and invertible, its inverse transformation A_t *also being absolutely continuous. The transformed process* $(K_s(A_t))$ *is again in* $\mathbb{L}^{1,2}$ *and such that*

$$\|(\int_0^1 \int_0^1 |D_r[K_s(A_t)]|^2 dr \, ds)^{1/2}\|_\infty \le \frac{1}{1 - c_K}, \quad 0 \le t \le 1.$$

Proposition 2.4.8 *(cf. Lemma 4.7 [10]) Let* $(K_s) \in \mathbb{L}^{1,2}$ *be as in Proposition 2.4.7. Then the process* $(D_r[K_s(A_t)])$ *has a version for which the function* $(t \mapsto D_r[K_s(A_t)] \in L^2(\Omega))$ *is continuous for all* $0 \le r, s \le 1$.

Proof (of Proposition 2.4.6): For each $0 < \varepsilon < 1$ define

$$T^\varepsilon\omega = \omega + (1 - \varepsilon)\int_0^{\cdot} K_s(\omega)ds, \quad \omega \in \Omega.$$

Obviously, $((1 - \varepsilon)K_s(\omega))$ satisfies the assumptions of the Propositions 2.4.5, 2.4.7 and 2.4.8. Denote the density of T^ε by L^ε, that of $A^\varepsilon = (T^\varepsilon)^{-1}$ by $\mathcal{L}^\varepsilon$, and recall that

$$\mathcal{L}^\varepsilon = |d_c(-(1 - \varepsilon)DK)| \exp\{-(1 - \varepsilon)\int_0^1 K_s dW_s - \frac{1}{2}(1 - \varepsilon)^2 \int_0^1 K_s^2 ds\},$$

where, with the notations

$$T_t^\varepsilon\omega = \omega + (1 - \varepsilon)\int_0^{t \wedge \cdot} K_s(\omega)ds \quad \text{and} \quad A_t^\varepsilon = (T_t^\varepsilon)^{-1},$$

it holds

$$|d_c(-(1-\varepsilon)DK)| = \exp\{-(1-\varepsilon)^2 \int_0^1 \int_0^s D_r K_s \cdot D_s[K_r(A_s^\varepsilon)](T_s^\varepsilon)dr\, ds\}.$$

By Proposition 2.1.5,

$$D_r[K_s(A_t^\varepsilon)] = (D_r K_s)(A_t^\varepsilon) - (1-\varepsilon)\int_0^t (D_u K_s)(A_t^\varepsilon)D_r[K_u(A_t^\varepsilon)]du,$$

$$\text{for a.e.}\quad 0 \le r,s \le 1.$$

By the Propositions 2.4.7 and 2.4.8 we may take the limit $t \to r$,

$$D_r[K_s(A_r^\varepsilon)] = (D_r K_s)(A_r^\varepsilon) - (1-\varepsilon)\int_0^r (D_u K_s)(A_r^\varepsilon)D_r[K_u(A_r^\varepsilon)]du,\quad \text{i.e.,}$$

$$D_r[K_s(A_r^\varepsilon)](T_r^\varepsilon) = D_r K_s - (1-\varepsilon)\int_0^r D_u K_s \cdot D_r[K_u(A_r^\varepsilon)](T_t^\varepsilon)dr,\quad \text{for a.e. } 0 \le s,r \le 1.$$

Hence, an easy estimate yields

$$\left(\int_0^1 \int_0^r |D_r[K_s(A_r^\varepsilon)](T_r^\varepsilon)|^2 dr\, ds\right)^{1/2} \le \frac{(\int_0^1 \int_0^r |D_r K_s|^2 ds\, dr)^{1/2}}{1 - (\int_0^1 \int_0^1 |D_r K_s|^2 ds\, dr)^{1/2}},\quad \text{a.e.,}$$

i.e.,

$$|d_c(-(1-\varepsilon)DK)| \le \exp\{\frac{(\int_0^1 \int_0^r |D_r K_s|^2 ds\, dr)^{1/2}(\int_0^1 \int_0^r |D_s K_r|^2 ds\, dr)^{1/2}}{1 - (\int_0^1 \int_0^1 |D_r K_s|^2 dr\, ds)^{1/2}}\}$$

$$= \exp\{\mathcal{E}\},\quad \text{a.e., for all}\quad 0 < \varepsilon < 1.$$

However, now we can show that the family of densities L^ε, $0 < \varepsilon < 1$, is uniformly integrable. For this it suffices to estimate $E[L^\varepsilon|\ln L^\varepsilon|]$ uniformly with respect to ε (cf. the proof of Lemma 2.2.9), which yields

$$E[L^\varepsilon|\ln L^\varepsilon|] = E[|\ln \mathcal{L}^\varepsilon|]$$

$$\le (E[\int_0^1 K_s^2 ds] + E[\int_0^1 \int_0^1 |D_r K_s|^2 dr\, ds])^{1/2} + \frac{1}{2}E[\int_0^1 K_s^2 ds] + E[\mathcal{E}],$$

where the right-hand side is independent of $0 < \varepsilon < 1$ and finite under the condition $\mathcal{E} \in L^1(\Omega)$.

Now, after checking the uniform integrability of the family of densities $L^\varepsilon = \frac{dP \circ [T^\varepsilon]^{-1}}{dP}$, we can derive immediately from Proposition 2.1.7 that T is absolutely continuous and

$L = \dfrac{dP \circ [T]^{-1}}{dP} = \sigma(L^1, L^\infty) - \lim_{\varepsilon \downarrow 0} L^\varepsilon.$ This completes the proof.

Remark: Enchev's approach [20], originally on the abstract Wiener space, uses the invertibility of the operator $I + DK(\omega) : L^2([0,1]) \to L^2([0,1])$ under the assumption $\int_0^1 \int_0^1 |D_r K_s(\omega)|^2 dr\, ds < 1$ for the Grohberg-Krein factorization

$$(I + DK(\omega))^{-1} = (I + V^+(\omega))(I + V^-(\omega)).$$

Here $V^+(\omega) = (V_{s,t}^+(\omega))$, $V^-(\omega) = (V_{s,t}^-(\omega))$ denote some elements of $L^2([0,1]^2)$ that have the property

$$\begin{aligned} V_{s,t}^+(\omega) &= 0 \quad \text{if} \quad s < t, \\ V_{s,t}^-(\omega) &= 0 \quad \text{if} \quad s > t, \end{aligned}$$

where I is the identity operator over $L^2([0,1])$.

This approach leads to the following main statement of Enchev.

Proposition 2.4.9 *Let* $(K_s) \in \mathbb{L}^{1,2}$ *be chosen such that*

$$\left(\int_0^1 \int_0^1 |D_r K_s|^2 dr\, ds \right)^{1/2} < 1, \qquad a.e.,$$

and put

$$L = \exp\{ -\int_0^1 K_s dW_s - \frac{1}{2} \int_0^1 K_s^2 ds - \int_0^1 \int_0^s D_r K_s \cdot V_{s,r}^+ dr\, ds \}.$$

If $E[L] = 1$, *then the transformation*

$$T\omega = \omega + \int_0^{\cdot} K_s(\omega) ds, \qquad \omega \in \Omega,$$

is absolutely continuous and $E[F(T)L] = E[F]$ *for all* $F \in L^\infty(\Omega)$.

The assumption $E[L] = 1$ of Proposition 2.4.9 is satisfied under the requirements for Proposition 2.4.6, and under the assumptions (i) and (ii) of Theorem 2.4.1 we have

$$\begin{aligned} V_{s,r}^+(\omega) &= D_s[K_r(A_s)](T_s\omega) I_{\{s \geq t\}}, \\ V_{s,r}^-(\omega) &= D_s[K_r(A_r)](T_r\omega) I_{\{s \leq t\}}, \quad a.e. \end{aligned}$$

2.5 Transformations defined by a shift adapted to some enlarged filtration

In this section we study the transformation $T\omega = \omega + \int\limits_0^\bullet K_s(\omega)ds$, $\omega \in \Omega$, of Ω into itself for a shift process $(K_s) \in L^{1,2}$ adapted to the natural filtration $(\mathcal{F}_s^W)$ of the coordinate process (W_s), enlarged by a Fréchet differentiable random variable G, and we will look for a density L changing the Wiener measure P into a probability measure $L \cdot P$ under which $((T\omega)_t = \omega_t + \int\limits_0^t K_s(\omega)ds)$ defines a Wiener process. In distinction to Section 2.4, the shift process (K_s) adapted to $(\mathcal{F}_s^W)$ enlarged by G will permit us to present a result about the absolute continuity of T that resembles the nonanticipative Girsanov Theorem more than Theorem 2.4.1.

Let $(K_s(\omega, x))$ be a measurable mapping of $[0,1] \times \Omega \times R^1$ into R^1 which belongs to $L^\infty(\Omega \times R^1, L^2([0,1]))$ and satisfies the following assumptions (C):

(C.i) For any $x \in R^1$, the process $K_\cdot(., x)$ is an $(\mathcal{F}_s^W)$-adapted element of $L^2([0,1], D^{1,\infty})$, and $(D_s K_t(\omega, x))$ belongs to $L^\infty(\Omega \times R^1, L^2([0,1]^2))$.

(C.ii) For almost any $(t, \omega) \in [0,1] \times \Omega$, the function $K_s(\omega, .)$ is in $C^1(R^1)$, for some $\delta > 0$ the process of the derivatives $(\frac{\partial}{\partial x}K_s(\omega, x))$ is in $L^\infty(\Omega \times R^1, L^{2+\delta}([0,1]))$, and moreover, for almost all $0 \le s \le 1$ and any $x \in R^1$, it holds that $\frac{\partial}{\partial x}K_s(., x) \in D^{1,2}$ and $x \mapsto \frac{\partial}{\partial x}K_s(\omega, x)$ is continuous, uniformly with respect to ω outside a subset of Ω of probability zero and not depending on s.

Since, due to (C.i), $K_\cdot(., x) \in L^2([0,1], D^{1,\infty})$, for any $x \in R^1$, we can utilize Theorem 2.2.1 in order to deduce the existence of a unique family of absolutely continuous and invertible transformations $T_t(., x)$, $0 \le t \le 1$, which satisfy the equation

$$T_t(\omega, x) = \omega + \int\limits_0^{t\wedge\cdot} K_s(T_s(\omega, x), x)ds, \quad \text{a.e.,} \quad 0 \le t \le 1.$$

Putting $T(\omega, x) = T_1(\omega, x)$ and taking into account the nonanticipativity of $(K_s(\omega, x))$ implies that $T(., x)$ is the strong solution of the nonanticipative equation

$$(2.5.1) \qquad T(\omega, x) = \omega + \int\limits_0^\bullet K_s(T(\omega, x), x)ds, \quad \text{a.e.}$$

The transformation $T(., x)$ is invertible, its inverse $A(., x)$ is given by

$$(2.5.2) \qquad A(\omega, x) = \omega - \int\limits_0^\bullet K_s(\omega, x)ds, \quad \text{a.e.}$$

Denote the densities of $T(., x)$ and $A(., x)$ by $L(x)$ and $\mathcal{L}(x)$, respectively. Clearly,

$$(2.5.3) \quad L(x) = \exp\{\int\limits_0^1 K_s(x)dW_s - \frac{1}{2}\int\limits_0^1 K_s(x)^2 ds\},$$

and

$$\mathcal{L}(x) \ (= L(T(x), x)^{-1}) = \exp\{-\int_0^1 K_s(T(x), x)dW_s - \frac{1}{2}\int_0^1 K_s(T(x), x)^2 ds\}.$$

For any $G \in \mathbb{D}^{1,\infty}$ we study the anticipative transformation

$$(2.5.4) \qquad\qquad T\omega = T(\omega, G(\omega)), \qquad \omega \in \Omega,$$

of Ω into itself, which is a strong solution of the anticipative stochastic differential equation

$$(T\omega)_t = \omega_t + \int_0^t K_s(T\omega, G(\omega))ds, \qquad a.e., \quad 0 \le t \le 1.$$

The main result we want to deduce in this section is the following:

Theorem 2.5.1 *Let* $(K_s(\omega, x))$ *be a process with the properties required in (C),* $A(x) \ (= A(., x))$ *the transformation introduced in (2.5.2), and* $\mathcal{L}(x)$ *its density. Moreover, let* $G \in \mathbb{D}^{1,\infty}$. *Suppose that, for some* $a > \|G\|_\infty$, *the random variable*

$$M(\omega, x) = \#\{y \in (-a, a) : y - G(A(\omega, y)) = x\}$$

is such that there exists the limit

$$M = L^1(\Omega) - \lim_{\varepsilon \downarrow 0} \int_{R^1} M(x)\mathcal{N}(0, \varepsilon)(dx).$$

Then, with the notation

$$(2.5.5) \qquad \mathcal{L}(\omega) = \mathcal{L}(\omega, G(\omega)) \cdot |1 + \int_0^1 (\partial_x K_s)(T\omega, G(\omega))D_s G(\omega)ds|$$

it holds
$$(2.5.6) \qquad\qquad E[F(T\omega)\mathcal{L}(\omega)] = E[F \cdot M], \quad \text{for all} \quad F \in L^\infty(\Omega).$$

Remark: In particular, if in addition to the assumptions of Theorem 2.5.1 the function $f(\omega, y) = y - G(A(\omega, y))$, $y \in R^1$, is pathwise injective, then $M(\omega, x) = 1$, for all $x \in R^1$, i.e., (2.5.6) takes the form

$$E[F(T\omega)\mathcal{L}(\omega)] = E[F], \qquad F \in L^\infty(\Omega).$$

Recall that the function $f(\omega, .) : R^1 \to R^1$ is surjective, for a.e. $\omega \in \Omega$, whenever G is essentially bounded.

The proof of this theorem is essentially based on the following lemma:

Lemma 2.5.2 *Under the assumptions of Theorem 2.5.1, the random functions* $g(x) = K(T(x), x)$ *and*

$$\ell(x) = \mathcal{L}(G + x) \cdot |1 + \int_0^1 (\frac{\partial}{\partial x} K_s)(T(G + x), G + x) D_s G ds|$$

are pathwise continuous mappings of R^1 *into* $L^2([0, 1])$ *and* R^1, *respectively, and, for each real* $c > 0$, *the random variable* $\sup_{|x| \leq c} \ell(x)$ *is in* $\bigcap_{p > 1} L^p(\Omega)$.

The correctness of this lemma will be shown after presenting of the proof of Theorem 2.5.1.

Proof of Theorem 2.5.1: Let F be a smooth Wiener functional of $\mathcal{S}$. Clearly, Lemma 2.5.2 allows us to apply the dominated convergence theorem in order to see that

$$(2.5.7) \quad E[F(T)\mathcal{L}] = \lim_{\varepsilon \downarrow 0} E[\int_{-\infty}^{+\infty} F(T(G + \varepsilon x))\ell(\varepsilon x) I_{(-a,a)}(G + \varepsilon x) \mathcal{N}(0, 1)(dx)].$$

However, as in the proof of Proposition 2.2.3, this makes it possible to eliminate the random variable G in the transformation $T(G + \varepsilon x)$ and to pass to the nonanticipative transformation $T(x)$. Therefore, denote the density of the Gaussian distribution $\mathcal{N}(0, \varepsilon^2)(dx)$ with respect to the Lebesgue measure dx by $\varrho_\varepsilon(x)$ and substitute y for $G + \varepsilon x$ in the right-hand side of (2.5.7). This yields:

$$(2.5.8) \quad E[\int_{-\infty}^{+\infty} F(T(G + \varepsilon x))\ell(\varepsilon x) I_{(-a,a)}(G + \varepsilon x) \mathcal{N}(0, 1)(dx)] =$$

$$= E[\int_{-\infty}^{+\infty} F(T(y))\mathcal{L}(y) \times$$

$$\times |1 + \int_0^1 (\frac{\partial}{\partial x} K_s)(T(y), y) \cdot D_s G ds| I_{(-a,a)}(y) \varrho_\varepsilon(y - G) dy].$$

Now, after changing the order of integration and expectation in the latter expression, we can apply the nonanticipative Girsanov theorem. Then, a renewed change of the order of integration and expectation shows that the expressions in (2.5.8) coincide with

$$(2.5.9) \quad E[F \int_{-\infty}^{+\infty} |1 + \int_0^1 (\frac{\partial}{\partial x} K_s)(y)(D_s G)(A(y)) ds| I_{(-a,a)}(y) \varrho_\varepsilon(y - G(A(y))) dy].$$

Taking into account that the function $f(\omega, y) = y - G(A(\omega, y))$ is pathwise absolutely continuous relative to y and has the derivative

$$f(\omega, y) = 1 + \int_0^1 (\frac{\partial}{\partial x} K_s)(\omega, y)(D_s G)(A(\omega, y)) ds,$$

the expression (2.5.9) takes the form

$$E[F \int\limits_{-\infty}^{+\infty} |\frac{\partial}{\partial y}f(y)|I_{(-a,a)}(y)\varrho_\varepsilon(f(y))dy].$$

The above expression is nothing else but

(2.5.10) $$E[F \int\limits_{-\infty}^{+\infty} M(x)\varrho_\varepsilon(x)dx],$$ cf. H. Federer [23].

Summarizing the computational steps (2.5.8) - (2.5.10) we obtain

(2.5.11) $$E[\int\limits_{-\infty}^{+\infty} F(T(G+\varepsilon x))\ell(\varepsilon x)I_{(-a,a)}(G+\varepsilon x)\mathcal{N}(0,1)(dx)]$$

$$= E[F \int M(x)\varrho_\varepsilon(x)dx], \quad \text{for all } 0 < \varepsilon < 1 \text{ and for all } F \in \mathcal{S}.$$

In virtue of the assumptions of the theorem, the limit of the right-hand side exists for $\varepsilon \downarrow 0$ and coincides with $E[FM]$. Hence, taking into account (2.5.7) we get the desired result $E[F(T)\mathcal{L}] = E[F \cdot M]$ for all $F \in \mathcal{S}$, and hence it holds for all $F \in L^\infty(\Omega)$, too.

It remains to prove Lemma 2.5.2, for which the following auxiliary statement is needed:

Lemma 2.5.3 *Suppose that the process $(K_s(\omega, x))$ satisfies (C). Then, for a.e. $(t, \omega) \in [0, 1] \times \Omega$, the function $K_t(T(\omega, x)x)$ is absolutely continuous relative to x, and (dropping ω) it holds*

(2.5.12) $$\frac{\partial}{\partial x}[K_t(T(x), x)] = (\frac{\partial}{\partial x}K_t)(T(x), x) + \int\limits_0^t (D_s K_t)(T(x), x)\frac{\partial}{\partial x}[K_s(T(x), x)]ds.$$

Moreover, there is a real $c_K > 0$ that bounds

$$\int\limits_0^1 |\frac{\partial}{\partial x}[K_t(T(x), x)]|^2 dt, \quad \text{for all } x \in R^1 \text{ and a.e. on } \Omega.$$

Proof: The desired result can be obtained by approximating $(K_t(\omega, x))$ by a sequence of $(\mathcal{F}_t^W)$-adapted processes $(K_t^n(\omega, x))$ of the form (2.3.4) which has been chosen such that

(i) for some real $C > 0$ it holds

$$\| \int\limits_0^1 |K_t^n(.,.)|^2 dt\|_{L^\infty(\Omega \times R^1)} \leq C,$$

$$\| \int\limits_0^1 \int\limits_0^1 |D_s K_t^n(.,.)|^2 ds\, dt\|_{L^\infty(\Omega \times R^1)} \leq C,$$

$$\| \int\limits_0^1 |\frac{\partial}{\partial x}K_t^n(.,.)|^2 dt\|_{L^\infty(\Omega \times R^1)} \leq C, \quad n = 1, 2, 3, \cdots$$

and

(ii) the processes $(K_t^n(\omega, x))$, $(\frac{\partial}{\partial x} K_t^n(\omega, x))$ and $(D_s K_t^n(\omega, x))$ converge to $(K_t(\omega, x))$, $(\frac{\partial}{\partial x} K_t(\omega, x))$ and $(D_s K_t(\omega, x))$ in $L^2([0,1] \times \Omega \times R^1, dt \times dP \times \mathcal{N}(0,1))$ and $L^2([0,1]^2 \times \Omega \times R^1, dsdt \times dP \times \mathcal{N}(0,1))$, respectively.

The existence of such a sequence can be shown by using the same arguments as for the derivation of Proposition 2.1.3 (cf. also Proposition 2.14 [8]).

Fix any natural n. Taking into account the special form of the $(\mathcal{F}_t^W)$-adapted smooth step process $(K_t^n(\omega, x))$

$$K_t^n(\omega, x) = \sum_{i=1}^{N} f_i(\omega(e_1), \ldots, \omega(e_{i-1}), x) e_i(t), \quad (t, \omega, x) \in [0,1] \times \Omega \times R^1,$$

for some $N \geq 1$, $0 = t_o < t_1 < \ldots < t_N = 1$, $e_i(t) = \frac{1}{\sqrt{t_i - t_{i-1}}} I_{[t_{i-1}, t_i]}(t)$,

and $f_i \in C_b^{\infty}(R^i)$, $i = 1, 2, \ldots, N$,

we see that, with the notations

$$\begin{aligned} g_1(\omega, x) &= \omega(e_1) + f_1(x) \\ g_2(\omega, x) &= \omega(e_2) + f_2(g_1(\omega, x), x) \\ &\vdots \\ g_N(\omega, x) &= \omega(e_N) + f_N(g_1(\omega, x), \ldots, g_{N-1}(\omega, x), x), \end{aligned}$$

the transformation

$$T^n(\omega, x) = \omega + \int_0^{\bullet} \sum_{i=1}^{N} f_i(g_1(\omega, x), \ldots, g_{i-1}(\omega, x), x) e_i(t) dt, \quad \omega \in \Omega,$$

solves the equation (2.5.1) for $(K_t^n(\omega, x))$. Clearly, the random variable

$$K_t^n(T^n(\omega, x), x) = \sum_{i=1}^{N} f_i(g_1(\omega, x), \ldots, g_{i-1}(\omega, x), x) e_i(t)$$

is $\mathcal{F}_t^W$-measurable, absolutely continuous with respect to x, and, as a straightforward computation shows,

(2.5.13)
$$\frac{\partial}{\partial x}[K_t^n(T^n(\omega, x), x)] = (\frac{\partial}{\partial x} K_t^n)(T^n(\omega, x), x)$$
$$+ \int_0^t (D_s K_t^n)(T^n(\omega, x), x) \frac{\partial}{\partial x}[K_s^n(T^n(\omega, x), x)] ds,$$

for all $(t, \omega, x) \in [0,1] \times \Omega \times R^1$.

For passing to the limit note that, according to the Girsanov theorem, the transformation $T^n(x)$ has the density

$$L^n(x) = \frac{dP \circ [T^n(x)]^{-1}}{dP} = \exp\{\int_0^1 K_t^n(x) dW_t - \frac{1}{2} \int_0^1 K_t^n(x)^2 dt\}$$

Obviously, in virtue of the assumptions (i) on the sequence $((K_t^n(\omega, x)))$, the family $\{L^n(x),\ x \in R^1,\ n = 1, 2, 3, \ldots\}$ of densities is uniformly integrable. This allows to repeat standard arguments of the proof of Theorem 2.2.1 in order to show that $(K_t^n(T^n(\omega, x), x))$ converges to $(K_t(T(\omega, x), x))$ in $L^2([0, 1] \times \Omega)$, for each $x \in R^1$. Once having done this, it is not hard to deduce from Proposition 2.1.8, using the assumptions (i) and (ii) for $(K_t^n(\omega, x))$, that

$$((\frac{\partial}{\partial x}K_t^n)(T^n(\omega, x), x)) \to ((\frac{\partial}{\partial x}K_t)(T(\omega, x), x)) \text{ in } L^2([0, 1] \times \Omega \times R^1, dt \times P \times \mathcal{N}(0, 1)),$$

and

$$((D_s K_t^n)(T^n(\omega, x), x)) \to ((D_s K_t)(T(\omega, x), x)) \text{ in } L^2([0, 1]^2 \times \Omega \times R^1, dsdt \times P \times \mathcal{N}(0, 1)).$$

In equation (2.5.13) we now pass to the limit in $L^2([0, 1] \times \Omega \times R^1, dt \times P \times \mathcal{N}(0, 1))$. Consequently, $K_t(T(\omega, x), x)$ is absolutely continuous with respect to x and satisfies (2.5.12) for a.e. $(t, \omega) \in [0, 1] \times \Omega$. However, having (2.5.12), assumption (C) guarantees the existence of a real $C_K > 0$ which bounds

$$\int_0^1 |\frac{\partial}{\partial x}[K_t(T(\omega, x), x)]|^2 dt \quad \text{for all } x \in R^1 \text{ and for a.e. } \omega \in \Omega.$$

Finally we present the proof of Lemma 2.5.2:

Proof of Lemma 2.5.2: By Lemma 2.5.3 we have the following pathwise estimation for $g_t(x) = K_t(T(x), x)$

$$(2.5.14) \quad \int_0^1 |g_t(x) - g_t(y)|^2 dt = \int_0^1 |\int_0^1 (\frac{\partial}{\partial x}g_t)(\theta x + (1 - \theta)y)d\theta|^2 dt \cdot |x - y|^2$$

$$\leq \|\int_0^1 |(\frac{\partial}{\partial x}g_t)(.)|^2 dt\|_{L^\infty(\Omega \times R^1)} \cdot |x - y|^2,$$

$$\text{for all } x, y \in R^1,$$

i.e., $g(x) = K(T(x), x)$ and $\int_0^1 K_t(T(x), x)^2 dt$ are pathwise continuous mappings of R^1 into $L^2([0, 1])$ and into R^1, respectively. Moreover, by (2.5.14) we can conclude from Kolmogorov's continuity criterion that $x \mapsto \int_0^1 K_t(T(x), x)dW_t$ is pathwise continuous, and, conseqently, that this is true for $\mathcal{L}(x)$ too. Taking into account (C.ii) and the pathwise continuity of g of R^1 into $L^2([0, 1])$, it is not hard to see that $(\frac{\partial}{\partial x}K_t)(T(x), x)$ is pathwise continuous relative to x. Thus, also $\ell(x)$ is pathwise continuous. It remains to show the L^p-integrability of $\sup_{|x| \leq c} \ell(x)$ for all real $p, c > 0$. Note that under the assumptions of Theorem 2.5.1, the random variable

$$\sup_{x \in R^1} |1 + \int_0^1 (\partial_x K_s)(T(x), x)D_s G ds|$$

is essentially bounded on $[0,1] \times \Omega$. On the other hand, for any $c, p > 1$, it holds

$$
\sup_{|x| \leq c} \mathcal{L}(x)^p \leq \exp\{p \int_0^1 K_s(T(0), 0) dW_s\}
$$

$$
+ p \int_{-c}^c \exp\{p \int_0^1 K_s(T(y), y) dW_s\} | \int_0^1 \frac{\partial}{\partial y}[K_s(T(y), y)] dW_s | dy\}
$$

$$
\leq \exp\{p \int_0^1 K_s(T(0), 0) dW_s\} + 2p \int_{-c}^c \exp\{2p \int_0^1 K_s(T(y), y) dW_s\} dy
$$

$$
+ 2p \int_{-c}^c | \int_0^1 \frac{\partial}{\partial y}[K_s(T(y), y)] dW_s |^2 dy.
$$

Taking into account the assumptions (C) for $(K_s(\omega, x))$ and Lemma 2.5.3, we see that the right-hand side of the above estimate is integrable. This completes the proof.

This section should not be concluded without a remark on the paper [15] which studies the same subject but by using another, in some sense more intuitive, but, also more technical approach, which is based on the quasi-sure analysis [1] of H. Airault and P. Malliavin. The basic idea of [15] is the following:

(For convenience) set $G = W_1$. In order to eliminate the anticipation $G = W_1$ in the transformation $T\omega = T(\omega, \omega_1)$, consider this transformation under the law P^x of the Brownian bridge $(\omega_t^x = \omega_t - t\omega_1 + tx)$ from 0 to x, establish a transformation formula for $E^x[F(T)\mathcal{L}]$, $F \in \mathcal{S}$, where $\mathcal{L}$ is defined by (2.5.5), and then derive the formula for $E[F(T)\mathcal{L}]$ by mixing $P = \int P^x W(0, 1)(dx)$.

3 Anticipative stochastic differential equations

The study of the anticipative stochastic differential equations is directly related to the theory of the stochastic integration of not necessarily adapted processes, i.e., the study of stochastic differential equations for which the solution is not required to be adapted to the driving Wiener process. For stochastic differential equations with the Stratonovich integral this has been done under a random initial condition as well as under boundary conditions by D. Ocone/E. Pardoux [37], [38], D. Nualart/E. Pardoux [34], C. Donati-Martin [20] and others. The case of stochastic differential equations has not been studied so thoroughly. An exclusion is given by the linear stochastic differential equation (with Skorohod integral),

$$(3.0.1) \qquad \begin{aligned} dX_t &= \sigma_t X_t W_t + b_t X_t dt, \\ X_o &= \eta. \end{aligned}$$

For a random initial condition η, but under the assumption of the deterministic processes (σ_s) and (b_s), this equation has been solved by Y. Shiota [42] and A.S. Ustunel [44] by means of the method of Wiener chaos decomposition. Another method, the method of the U-transformation of the Hida calculus for solving such equations with deterministic processes (σ_s) and (b_s) has been presented recently by several authors, e.g., by H.H. Kuo, B. Øksendal and J. Potthoff.

In distinction to these methods we use Girsanov transformations as the main tool for solving equation (3.0.1), where the results on the anticipative Girsanov transformations allow us to consider possibly anticipating processes (σ_s) and (b_s). This will be done in Section 3.2. Section 3.3 is devoted to Skorohod stochastic differential equations

$$(3.0.2) \qquad \begin{aligned} dX_t &= a(X_t)dW_t + b(X_t)dt \\ X_o &= \eta, \end{aligned}$$

with $C^2(R^1)$-functions a and b. Setting $\eta = f(W_1)$ in (3.0.2) and looking for a solution of the form $X_t = X_t(x)|_{x=W_1}$, we have to solve the following nonanticipative first order stochastic partial differential equation with Itô integral:

$$(3.0.3) \qquad \begin{aligned} dX_t(x) + a'(X_t(x))\partial_x X_t(x)dt &= a(X_t(x))dW_t + b(X_t(x))dt, \\ X_o(x) &= f(x). \end{aligned}$$

H. Kunita [27] has shown that, in general, there exists only a local solution $\{X_t(x), 0 \le t < T(x), x \in R^1\}$ whose life time $T(x) > 0$ is some stopping time that is lower semicontinuous with respect to x; only for some special $a(x)$, e.g., $a(x) = a \cdot x$, there is a global solution $\{X_t(x), 0 \le t \le 1, x \in R^1\}$. This is the reason why, in general, also equation (3.0.2) can have local solutions only, whereas semi-linear stochastic differential equations, cf. [4], and, in particular, (3.0.1) have a global solution.

The chapter is organized as follows: In Section 3.1 we complete the review on the anticipative stochastic calculus of Chapter 1 by additional notions (e.g., the Skorohod integral on certain subsets of the Wiener space), which we will need in order to introduce the concept of a (local) solution of an anticipative stochastic differential equation with Skorohod integral. Then, Section 3.2 is devoted to linear Skorohod stochastic differential equations, and in Section 3.3 nonlinear Skorohod equations of type (3.0.3) are studied. Finally, in Section 3.4 linear Skorohod equations with boundary conditions are considered. While the Sections 3.1 - 3.3 are based on the papers [11] and [12], Section 3.4 follows a common work with D. Nualart [16] on semi-linear stochastic differential equations with boundary conditions.

3.1 Basic notions. Skorohod integral on balls

We will use $(\Omega, \mathcal{F}, P)$ again to denote the Wiener space, in particular $\Omega = C_o([0,1])$, and (W_s) denotes the coordinate process on Ω.

A random variable $F : \Omega \to R^1$ is said to be an element of $\mathcal{S}^*$ if it belongs to $L^*(\Omega) = \bigcap_{p>1} L^p(\Omega)$ and if there is an $L^2([0,1])$-valued random variable $(D_s F) \in L^*([0,1], L^2([0,1])) = \bigcap_{p>1} L^p(\Omega, L^2([0,1]))$ such that, for any $\omega \in \Omega$,

$$F\left(\omega + \int_0^{\bullet} h_s ds\right) = F(\omega) + \int_0^1 D_s F(\omega) h_s ds + o(|h|_2),$$

as $|h|_2 \to 0$, $h \in L^2([0,1])$.

In particular, all smooth Wiener functionals of $\mathcal{S}$ belong to $\mathcal{S}^*$, all elements of $\mathcal{S}^*$ are Fréchet differentiable. Generalizing the Fréchet differentiability, a random variable $F : \Omega \to R^1$ is called Malliavin differentiable if there exists an $L^2([0,1])$-valued random variable $(D_s F)$ such that, for any $h \in L^2([0,1])$ and $\delta > 0$,

$$\lim_{\varepsilon \downarrow 0} P\left\{\omega : |\frac{1}{\varepsilon}(F(\omega + \varepsilon \int_0^{\bullet} h_s ds) - F(\omega)) - \int_0^1 D_s F(\omega) h_s ds| > \delta\right\} = 0.$$

The set of the Malliavin differentiable random variables will be denoted by $\mathbb{D}^{1,0}$. Obviously, with the notation $\mathbb{D}^{1,*} = \bigcap_{p>1} \mathbb{D}^{1,p}$ we have

$$\mathbb{D}^{1,*} \subset \mathbb{D}^{1,p} \subset \mathbb{D}^{1,0}, \qquad p > 1.$$

Lemma 3.1.1 *For any $h \in L^2([0,1])$ and $\theta_h \omega = \omega + \int_0^{\bullet} h_s ds$, $\omega \in \Omega$, the spaces $\mathbb{D}^{1,0}$ and $\mathbb{D}^{1,*}$ are invariant relative to the transformation θ_h, i.e., for any $F \in \mathbb{D}^{1,0}$ ($\mathbb{D}^{1,*}$), also $F(\theta_h)$ belongs to $\mathbb{D}^{1,0}$ ($\mathbb{D}^{1,*}$), and*

$$D_s[F(\theta_h)] = (D_s F)(\theta_h), \qquad a.e.$$

Lemma 3.1.2 *(cf. [35]) The random variables* $M = \max_t W_t$, $m = \min_t W_t$ *and* $\|W\| = \max_t |W_t|$ *are in* $\mathbb{D}^{1,*}$ *and, with the notations* $\tau_1 = \min\{t \in [0,1] : W_t = M\}$ *and* $\tau_s = \min\{t \in [0,1] : W_t = m\}$, *it holds*

$$D_s M = I_{[0,\tau_1]}(s), \qquad D_s m = I_{[0,\tau_2]}(s), \quad and$$
$$D_s[\|W\|] = I\{M > -m\}I_{[0,\tau_1]}(s) - I\{M < -m\}I_{[0,\tau_2]}(s), \quad a.e.$$

In particular, $(D_s M)$, $(D_s m)$ *and* $(D_s[\|W\|])$ *are bounded by 1.*

By virtue of the Lemmata 3.1.1 and 3.1.2 the random variable $\|W(\theta_h)\|$ belongs to $\mathbb{D}^{1,*}$ for any $h \in L^2([0,1])$, and the derivative is bounded by 1.

Finally, in addition to the space $\widetilde{\mathbb{D}^{1,\infty}}$ of all $F \in L^\infty(\Omega) \cap \mathbb{D}^{1,*}$ with $(D_s F) \in L^\infty([0,1] \times \Omega)$, for any open ball $A = B_r(h)$ with radius $r > 0$ around $\int_0^\bullet h_s ds$, $h \in L^2([0,1])$ we introduce the space $\widetilde{\mathbb{D}^{1,\infty}}(A)$ of all $F \in \mathbb{D}^{1,\infty}$ whose support is in A and has a strictly positive distance to the boundary of A.

Lemma 3.1.3 *If, for any* $G \in L^1(A)$, *all* $F \in \mathbb{D}^{1,\infty}(A)$ *satisfy the relation*

$$(3.1.1) \qquad\qquad \int_A FG dP = 0,$$

then $G = 0$ *a.e. on* A.

Proof: Let $A = B_r(h)$ and (H_m) be a sequence of Fréchet differentiable random variables approximating sign $G \cdot I_A$ in $L^2(\Omega)$.

Then, for any $\varphi \in C_o^\infty(R^1)$ with supp $\varphi \subset (-r, r)$, the sequence $\{((H_m \wedge 1) \vee (-1)) \cdot \varphi(\|W(\theta_h)\|), m = 1, 2, 3, \ldots\}$ is in $\mathbb{D}^{1,\infty}(A)$, is bounded in $L^\infty(\Omega)$ and approximates $G \cdot I_A$ in $L^2(\Omega)$. Hence, due to (3.1.1),

$$\int_A |G| \cdot \varphi(\|W(\theta_h)\|) dP$$

for all $\varphi \in C_o^\infty(R^1)$ with supp $\varphi \subset (-r, r)$. This provides the result.

Now we can introduce the Skorohod integral $(\delta_A, \text{Dom } \delta_A)$ on A as the adjoint of $(D, \widetilde{\mathbb{D}^{1,\infty}}(A))$:

Definition: A process $(u_s) \in L^1([0,1] \times A)$ is said to be Skorohod integrable on $A = B_r(h)$ if there exists a $\delta_A(u) \in L^1(A)$ such that

$$(3.1.2) \qquad \int_A F \delta_A(u) dP = \int_A (\int_0^1 (D_s F \cdot u_s ds) dP, \quad \text{for all } F \in \mathbb{D}^{1,\infty}(A).$$

In this case we write $(u_s) \in \text{Dom } \delta_A$ and call the unique element $\delta_A(u) \in L^1(A)$ the Skorohod integral of (u_s) on A.

Note that $(\delta_\Omega, \text{Dom } \delta_\Omega) = (\delta, \text{Dom } \delta)$. Moreover, the relation of $(\delta_A, \text{Dom } \delta_A)$ to the Skorohod integral δ introduced in Chapter 1 will be elucidated by the following obvious lemma:

Lemma 3.1.4 *If A and B are arbitrary balls in Ω with $A \subset B$, then every process $(u_s) \in Dom\,\delta_B$ considered as an element of $L^1([0,1] \times A)$ is also in $Dom\,\delta_A$, and*

$$\delta_A(u) = \delta_B(u), \quad a.e. \text{ on } A.$$

Lemma 3.1.4 allows us to write $\int_0^t u_s dW_s$ for $\delta_A(uI_{[0,t]})$ whenever the process $(u_s I_{[0,t]}(s))$ belongs to $\mathrm{Dom}\,\delta_A$ for some $t \in [0,1]$ and a ball $A \subset \Omega$. Moreover, note that δ_A obeys the local property, i.e., for any $(u_s) \in \mathrm{Dom}\,\delta_A$,

$$\delta_A(u) = 0 \quad \text{a.e. on any ball } B \subset \{\int_0^1 |u_s|^2 ds = 0\} \ (\subset A).$$

After this preparation we can introduce the notion of a (local) solution of a Skorohod stochastic differential equation. Fix any $h \in L^2([0,1])$, $r > 0$ (possibly infinity) and $0 \le t \le 1$, and set $A = B_r(h)$.

Definition: Let $X_o \in L^1(A)$, $(a_s(x))$ and $(b_s(x))$ be in $L^1([0,t] \times A \times R^1)$. A process $(X_s) \in L^1([0,t] \times A)$ is called a solution of the Skorohod equation

$$(3.1.3) \qquad X_s = X_o + \int_0^s a_r(X_r)dW_r + \int_0^s b_r(X_r)dr \quad \text{a.e. on } A, \ s \in [0,t],$$

if

(i) $X_s \in L^1(A)$, $s \in [0,t]$, $(a_r(X_r))$ and $(b_r(X_r))$ are in $L^1([0,t] \times A)$,

(ii) $(a_r(X_r)I_{[0,s]}(r)) \in \mathrm{Dom}\,\delta_A$, for all $s \in [0,t]$, and

(iii) equation (3.1.3) holds, i.e., for any $F \in \widetilde{D^{1,\infty}}(A)$, $s \in [0,t]$, we have

$$\int_A (\int_0^s a_r(X_r)D_r F dr)dP = \int_A \{X_s - X_o - \int_0^s b_r(X_r)dr\}F dP$$

Definition: A process $(X_s) = \{X_s(\omega), 0 \le s < \tau(\omega)\}$ with life time τ (strictly positive random variable) is called a local solution of equation (3.1.3) if there is a $t \in (0,1)$ for any bounded ball $A \subset \Omega$ such that

(i) $t < \tau(\omega)$, $\omega \in A$,

(ii) $\int_A (\int_0^t |X_s|ds)dP < \infty$, and

(iii) (X_s) considered as an element of $L^1([0,t] \times A)$ is a solution of (3.1.3) on A.

A process $(X_s) = \{X_s(\omega), 0 \le s \le 1\} \in L^1([0,1] \times \Omega)$ is called a global solution of (3.1.3) if it solves (3.1.3) for $A = \Omega$ and $t = 1$.

3.2   Linear Skorohod stochastic differential equations

The main part of this section is devoted to the study of linear stochastic differential
equations

$$(3.2.1) \qquad X_t = \eta + \int_0^t \sigma_s X_s dW_s + \int_0^t b_s X_s ds, \quad 0 \le t \le 1,$$

where the coefficient (σ_s) is supposed to belong to $L^2([0,1], I\!\!D^{1,\infty})$, but the drift coefficient (b_s) and the initial value η are only assumed to be bounded. The main result of
this section is the following:

Theorem 3.2.1 *Suppose* $(\sigma_s) \in L^2([0,1], I\!\!D^{1,\infty})$, $(b_s) \in L^1([0,1], L^\infty(\Omega))$ *and* $\eta \in$
$L^\infty(\Omega)$. *Denote by* $\{T_t, t \in [0,1]\}$ *the family of absolutely continuous transformations
associated to* (σ_s) *by Theorem 2.2.1, let* A_t *be the inverse to* T_t, *and* L_t *be the density
of* T_t. *Then the process* (X_t) *defined by*

$$(3.2.2) \qquad X_t = \eta(A_t) \exp\{\int_0^t b_s(T_s A_t) ds\} L_t, \quad 0 \le t \le 1,$$

belongs to $L^1([0,1] \times \Omega)$ *and is a global solution of equation (3.2.1). Conversely, if*
$(Y_s) \in L^1([0,1] \times \Omega)$ *is a global solution of (3.2.1), and if, moreover,* (σ_s) *and* (b_s) *are
in* $L^\infty([0,1] \times \Omega)$ *and* $(D_r \sigma_s)$ *in* $L^\infty([0,1]^2 \times \Omega)$, *then* Y_t *is of the form (3.2.2) for a.e.*
$0 \le t \le 1$.

After this global statement, equation (3.2.1) will be studied for processes $(\sigma_s), (b_s) \in$
$L^\infty([0,t] \times A)$ and an initial value $\eta \in L^1(A)$ defined on a ball $A = B_r(h)$ with radius
$r > 0$ around any $\int_0^\bullet h_s ds$, $h \in L^2([0,1])$. The corresponding result will be used to
show the uniqueness of local solutions of nonlinear stochastic differential equations
with anticipation. Finally, again for linear stochastic differential equations, we will
study diffusion coefficients (σ_s) being not smooth in the sense of the Malliavin calculus
but adapted to some enlargement of the natural filtration of the Wiener process.
We now prove the first part of Theroem 3.2.1, i.e. the existence of a global solution.
The proof of its uniqueness will be completely covered by the uniqueness of a local
solution on any $[0,t] \times B_r(h)$, $h \in L^2([0,1])$, $0 < r \le +\infty$, which will be proved later.

Proof of Theorem 3.2.1 (Existence): Let the process (X_s) be defined by (3.2.2). Then,
clearly, (X_s) belongs to $L^1([0,1] \times \Omega)$, and, moreover, it holds that $X_t \in L^1(\Omega)$, $0 \le$
$t \le 1$. Thus, it remains to show that $(\sigma_s X_s I_{[0,t]}(s)) \in \text{Dom}\,\delta$, $0 \le t \le 1$, and (X_s)
verifies (3.2.2).
Assume that G is an element of the set of smooth Wiener functionals $\mathcal{S}$ and let us
compute $E[\int_0^t \sigma_s X_s D_s G ds]$. By (3.2.2) and the transformation of the measure $P \to$
$P \circ [T_s]^{-1} = L_s P$ we see that this expression is equal to

$$(3.2.3) \qquad E[\int_0^t \sigma_s \eta(A_s) L_s \exp\{\int_0^s b_r(T_r A_s) dr\} D_s G ds]$$

$$= E[\int_0^t \sigma_s(T_s)\eta \exp\{\int_0^s b_r(T_r)dr\}(D_sG)(T_s)ds].$$

Due to Lemma 2.2.7 (i), $\frac{d}{ds}G(T_s) = \sigma_s(T_s)(D_sG)(T_s)$. Therefore, integrating by parts in (3.2.3) and a renewed transformation of measures provide:

$$E[\int_0^t \eta \exp\{\int_0^s b_r(T_r)dr\}\frac{d}{ds}G(T_s)ds]$$

$$= E[\eta \exp\{\int_0^t b_s(T_s)ds\}G(T_t) - \eta G - \int_0^t \eta b_s(T_s)\exp\{\int_0^s b_r(T_r)dr\}G(T_s)ds]$$

$$= E[\eta(A_t)\exp\{\int_0^t b_s(T_sA_t)ds\}L_tG] - E[\eta G]$$

$$- E[\int_0^t \eta(A_s)b_s \exp\{\int_0^s b_r(T_rA_s)dr\}L_sG ds]$$

$$= E[X_tG] - E[\eta G] - E[\int_0^t b_sX_sdsG].$$

Clearly, the random variable $X_t - \eta - \int_0^t b_sX_sds$ is integrable, and we can deduce that $(\sigma_sX_sI_{[0,t]}(s))$ belongs to the domain of δ and

$$\int_0^t \sigma_sX_sdW_s = X_t - \eta - \int_0^t b_sX_sds, \qquad \text{a.e.,}$$

i.e., (X_s) is a global solution of (3.2.1).

Remark: Taking into account that $\mathcal{L}_t = \frac{dP \circ [A_t]^{-1}}{dP}$ satisfies the relation $L_t = \mathcal{L}_t(A_t)^{-1}$, Theorem 2.2.1, (2.2.2) provides

$$L_t = \exp\{\int_0^t \sigma_r(T_rA_t)dW_r - \frac{1}{2}\int_0^t \sigma_r(T_rA_t)^2 dr - \int_0^t \int_r^t (D_u\sigma_r)(T_rA_t)D_r[\sigma_u(T_uA_t)]dudr\}.$$

The global statement presented by Theorem 3.2.1 can be generalized to a local one. Fix any ball $A = B_r(h)$ with radius $r > 0$ around any $\int_0^{\bullet} h_sds$, $h \in L^2([0,1])$, and any $0 \le t \le 1$.

Theorem 3.2.2 *Let $(\tilde{\sigma}_s)$, $(\tilde{b}_s)$ be in $L^\infty([0,t] \times A)$ and $\tilde{\eta} \in L^\infty(A)$, and suppose that there is a process $(\sigma_s) \in L^\infty([0,t], \widetilde{D^{1,\infty}})$ which coincides with $(\tilde{\sigma}_s)$ a.e. on $A = B_r(h)$. Then there exists a solution $(\tilde{X}_s) \in L^1([0,t] \times A)$ of the linear equation*

$$(3.2.4) \qquad \tilde{X}_s = \tilde{\eta} + \int_0^s \tilde{\sigma}_r \tilde{X}_r dW_r + \int_0^s \tilde{b}_r \tilde{X}_r dr \quad a.e.\ on\ A,\ 0 \le s \le t.$$

This solution is unique in $B_{r-3\varphi(t)}(h)$, where

$$\varphi(t) = \int_0^t \|\sigma_s\|_\infty ds,$$

and if we set $\eta = \tilde{\eta}$ on A, $\eta = 0$ outside A, $(b_s) = (\tilde{b}_s)$ on $[0,t] \times A$, and $(b_s) = 0$ outside $[0,t] \times A$, then $(\tilde{X}_s)$ is given on $B_{r-3\varphi(t)}(h)$ by (3.2.2).

Proof (Existence): Due to the part of Theorem 3.2.1 that is already proved, the process (X_s) given by (3.2.2) is a solution of (3.2.1), i.e., $(X_s) \in L^1([0,t] \times \Omega)$ with $X_s \in L^1(\Omega)$, $(\sigma_r X_r I_{[0,s]}(r)) \in \mathrm{Dom}\,\delta$ and

$$\delta(\sigma X I_{[0,s]}) = X_s - X_o - \int_0^s b_r X_r dr \quad a.e.,\ 0 \le s \le t.$$

Then it follows from Lemma 3.1.4 that the restriction $(\tilde{X}_s)$ of (X_s) to A is in $L^1([0,t] \times A)$, $\tilde{X}_s \in L^1(A)$, $(\tilde{\sigma}_r \tilde{X}_r I_{[0,s]}(r)) \in \mathrm{Dom}\,\delta_A$ and

$$\delta_A(\tilde{\sigma} \tilde{X} I_{[0,s]}) = \tilde{X}_s - \tilde{X}_o - \int_0^s \tilde{b}_r \tilde{X}_r dr \quad a.e.\ on\ A,\ for\ all\ 0 \le s \le t,$$

i.e., $(\tilde{X}_s)$ is a solution of (3.2.4).

The proof of the uniqueness requires to recall Proposition 2.1.3 and Lemma 2.2.7 as well as a fact from the proof of Theorem 2.2.1. For the reader's convenience we present the necessary facts, but now already in a form which allows us to apply them immediately to the proof.

Lemma 3.2.3 *Let $(\sigma_s) \in L^\infty([0,t], \widetilde{D^{1,\infty}})$. Then there exists a sequence of smooth step processes $((\sigma_s^n)) \subset \mathbb{L}^S$ with the following properties:*

(i) (σ_s^n) converges to (σ_s) in $L^2([0,t], D^{1,2})$, and

(ii) $\|\sigma^n\|_{L^\infty([0,t] \times \Omega)} \le \|\sigma\|_{L^\infty([0,t] \times \Omega)}$,

$$\|D\sigma^n\|_{L^\infty([0,t]^2 \times \Omega)} \le 1 + \|D\sigma\|_{L^\infty([0,t]^2 \times \Omega)}, \quad n = 1,2,3,\ldots$$

For a given process $(\sigma_s) \in L^\infty([0,t], \widetilde{D^{1,\infty}})$ fix such a sequence $((\sigma_s^n))$, and denote by $A_s^n : \Omega \to \Omega$ the inverse transformation of $T_s^n : \Omega \to \Omega$ associated to (σ_s^n) by Theorem 2.2.1.

In the same manner associate the transformations T_s and A_s to the process (σ_s). Then we can state:

Lemma 3.2.4 *For any $F \in S$ and with the above notations we have:*

(i) $(F(A_s^n)) \in L^\infty([0,t], D^{1,\infty})$ *has been chosen such that $F(A_s^n) \in S$ for all $0 \le s \le t$.*

(ii) *The mapping $r \mapsto D_s[F(A_r^n)]$ is continuous for a.e. $(s,\omega) \in [0,t] \times \Omega$, and we can set*

$$D_s[F(A_s^n)] = \lim_{r \to s} D_s[F(A_r^n)], \quad s \in [0,t].$$

The process $(D_s[F(A_s^n)])$ belongs to $L^\infty([0,t], \widetilde{D^{1,\infty}})$, and

$$(3.2.5) \quad \|D_s[F(A_s^n)]\|_\infty \le \|DF\|_{L^\infty([0,t] \times \Omega)} \exp\{2(1 + \|D\sigma\|_{L^\infty([0,t]^2 \times \Omega)})\},$$
$$\text{for all} \quad 0 \le s \le t.$$

(iii) *The mapping $r \mapsto F(A_r^n)$ is a.e. absolutely continuous with respect to the Lebesgue measure*

$$(3.2.6) \qquad \frac{d}{ds}F(A_s^n) = -\sigma_s^n D_s[F(A_s^n)], \quad a.e.$$

Lemma 3.2.5 *For any $F \in S$, it holds*

$$F(A) = L^2(\Omega) - \lim_{n \to \infty} F(A_s^n), \qquad s \in [0,t].$$

Now we can show the uniqueness of the solution of (3.2.4) by means of the notations introduced above:

Proof of Theorem 3.2.2 (Uniqueness): Fix any $\varepsilon > 0$ and any $C_o^\infty(R^1)$-function ψ having all its values between 0 and 1, taking the value 1 in the interval $[-(r - \varphi(t)) + \varepsilon, (r - \varphi(t)) - \varepsilon]$, and having its support inside $[-(r - \varphi(t)) + \frac{\varepsilon}{2}, (r - \varphi(t)) - \frac{\varepsilon}{2}]$. Set $H = \psi(\|W(\theta_h)\|$ and let $F \in S$. Then, for any $0 \le s \le t$ and any natural n, the random variable $HF(A_s^n)$ belongs to $\widetilde{D^{1,\infty}}(A)$, and (3.2.4) provides

$$(3.2.7) \quad \int_A \tilde{X}_s H F(A_s^n) dP = \int_A \tilde{\eta} H F(A_s^n) dP + \int_A \left(\int_0^s \tilde{\sigma}_r \tilde{X}_r D_r[H F(A_s^n)] dr \right) dP$$

$$+ \int_A \left(\int_0^s \tilde{b}_r \tilde{X}_r H F(A_s^n) dr \right) dP.$$

By Lemma 3.2.4 we have

$$F(A_s^n) = F - \int_0^s \sigma_v^n D_v[F(A_v^n)] dv,$$

and, for $u \leq s$,

$$D_u[HF(A_s^n)] = D_u[HF(A_u^n)] - \int_u^s D_u[\sigma_v^n H D_v[F(A_v^n)]]dv.$$

Thus, the right-hand side of (3.2.7) reads now

$$(3.2.8) \quad \int_A \tilde{\eta} HF dP + \int_A (\int_0^s \tilde{\sigma}_u \tilde{X}_u D_u[HF(A_u^n)]du)dP + \int_A (\int_0^s \tilde{b}_u \tilde{X}_u HF(A_u^n)du)dP$$

$$- \int_0^s \{\int_A \tilde{\eta} H \sigma_v^n D_v[F(A_v^n)]dP + \int_A \int_0^v \tilde{\sigma}_u \tilde{X}_u D_u[H\sigma_v^n D_v[F(A_v^n)]]dudP$$

$$+ \int_A \int_0^v \tilde{b}_u \tilde{X}_u H \sigma_v^n D_v[F(A_v^n)]dudP\}dv.$$

Applying (3.2.4) to the random variable $H\sigma_v^n D_v[F(A_v^n)] \in \widetilde{I\!D^{1,\infty}}(A)$, we see that $\int_A(\int_0^s \tilde{X}_u \sigma_u^n H D_u[F(A_u^n)]dudP$ is equal to the integral in the last two lines of (3.2.8), i.e.,

$$\int_A \tilde{X}_s HF(A_s^n)dP - \int_A \tilde{\eta} HF dP - \int_A (\int_0^s \tilde{b}_u \tilde{X}_u HF(A_u^n)du)dP$$

$$= \int_A (\int_0^s (\tilde{\sigma}_u - \sigma_u^n)\tilde{X}_u D_u[HF(A_u^n)]du)dP + \int_A (\int_0^s \sigma_u^n \tilde{X}_u F(A_u^n)D_u H du)dP.$$

Since $F \in \mathcal{S}$ is bounded, due to Lemma 3.2.5 we can to apply the dominating convergence theorem to the left-hand side of this equation. Then the limit of the left-hand side equals

$$\int_A \tilde{X}_s HF(A_s)dP - \int_A \tilde{\eta} HF dP - \int_A (\int_0^s \tilde{b}_u \tilde{X}_u HF(A_u)du)dP.$$

Taking into account that $(\tilde{\sigma}_u)$ coincides with (σ_u) on A, we derive from the Lemmata 3.2.3 (ii) and 3.2.4 (ii) that the assumptions required for the dominated convergence theorem are satisfied for the first integral on the right-hand side, too. Hence, Lemma 3.2.3 (i) implies that

$$\lim_{n\to\infty} \int_A (\int_0^s (\tilde{\sigma}_u - \sigma_u^n)\tilde{X}_u D_u[HF(A_u^n)]du)dP = 0.$$

The same arguments provide

$$\lim_{n\to\infty} \int_A (\int_0^s \sigma_u^n \tilde{X}_u F(A_u^n)D_u H du)dP = \int_A (\int_0^s \tilde{\sigma}_u \tilde{X}_u F(A_u)D_u H du)dP.$$

Consequently,

$$(3.2.9) \qquad \int_A \tilde{X}_s H F(A_s) dP = \int_A \tilde{\eta} H F dP + \int_A (\int_0^s \tilde{b}_u \tilde{X}_u H F(A_u) du) dP$$

$$+ \int_A (\int_0^s \tilde{\sigma}_u \tilde{X}_u F(A_u) D_u H du) dP.$$

Since supp $H \subset B_{r-\varphi(t)}(h)$ and the transformation $A_s \omega = \omega - \int_0^{s \wedge \cdot} \sigma_u(A_{u,s} \omega) du$ $(A_{u,s} = T_u \circ A_s)$ maps $B_{r-\varphi(t)}(h)$ into A, a Girsanov transformation in (3.2.9) yields

$$(3.2.10) \quad \int_A \tilde{X}_s(T_s) H(T_s) F \mathcal{L}_s dP = \int_A \tilde{\eta} H F dP + \int_A (\int_0^s \tilde{b}_u(T_u) \tilde{X}_u(T_u) H(T_u) F \mathcal{L}_u du) dP$$

$$+ \int_A (\int_0^s \tilde{\sigma}_u(T_u) \tilde{X}_u(T_u) (D_u H)(T_u) \mathcal{L}_u F du) dP.$$

Here T_s denotes the inverse transformation to A_s, and $\mathcal{L}_s$ is the density of A_s. All integrals

$$(3.2.11) \qquad \int_A |H(T_s) \tilde{X}_s(T_s) \mathcal{L}_s| dP = \int_A |\tilde{X}_s| H dP$$

$$\int_A (\int_0^s |H(T_u) \tilde{b}_u(T_u) \tilde{X}_u(T_u) \mathcal{L}_u| du) dP = \int_A (\int_0^s |\tilde{b}_u \tilde{X}_u| du \cdot H) dP$$

$$\text{and}$$

$$\int_A (\int_0^s |(D_u H)(T_u) \tilde{\sigma}_u(T_u) \tilde{X}_u(T_u) \mathcal{L}_u| du) dP = \int_A (\int_0^s |\tilde{\sigma}_u \tilde{X}_u D_u H| du) dP$$

are finite. Hence, (3.2.10) does not only hold for all $F \in \mathcal{S}$, but also for all $F \in L^\infty(\Omega)$. Let now F be any element of $L^\infty(\Omega)$ with support in $B_{r-2\varphi(t)-\varepsilon}(h)$. In this ball $H(T_s)$ takes only the value 1, and the local property of D,

$$D_u H = 0 \quad \text{a.e. on} \quad \{H = 1\} \quad (\supset B_{r-\varphi(t)-\varepsilon}(h)),$$

implies

$$(D_u H)(T_u) = 0 \quad \text{a.e. on} \quad B_{r-2\varphi(t)-\varepsilon}(h).$$

Hence, we can conclude from (3.2.10) and (3.2.11) that, for any $0 \le s \le t$,

$$\tilde{X}_s(T_s) \mathcal{L}_s = \tilde{\eta} + \int_0^s \tilde{b}_u(T_u) \tilde{X}_u \mathcal{L}_u du \quad \text{a.e. on} \quad B_{r-2\varphi(t)-\varepsilon}(h),$$

i.e.,

$$\tilde{X}_s(T_s) \mathcal{L}_s = \tilde{\eta} \exp\{\int_0^s \tilde{b}_u(T_u) du\}.$$

This shows that the process $(\tilde{X}_s(T_s))$ is unique in $B_{r-2\varphi(t)-\varepsilon}(h)$, and so is (X_s) in $B_{r-3\varphi(t)-\varepsilon}(h)$. Since $\varepsilon > 0$ is arbitrary, (X_s) must be unique in the ball $B_{r-3\varphi(t)}(h)$, too. This completes the proof.

The uniqueness of the solution of a linear Skorohod stochastic differential equation in a ball in Ω allows us to deduce the uniqueness of the solution also for a nonlinear Skorohod stochastic differential equation. Let $r > 0$, $h \in L^2([0,1])$ and $A = B_r(h)$, and denote by $D_t(A)$ the set of all $(u_s) \in L^1([0,t] \times A)$ which possess an extension $(\tilde{u}_s) \in L^1([0,t] \times \Omega)$ such that $\tilde{u}_s \in \mathbb{D}^{1,0}$ for all $0 \le s \le t$ and $(D_r\tilde{u}_s) \in L^\infty([0,1] \times [0,t] \times \Omega)$.

Theorem 3.2.6 *Let $(a_s(x))$ and $(b_s(x))$ be elements of $L^1([0,t] \times A \times R^1)$ such that $a_s(\omega,.)$ and $b_s(\omega,.)$ belong to $C^2(R^1)$ a.e., and both $(\frac{\partial}{\partial x}a_s(x))$ and $(\frac{\partial}{\partial x}b_s(x))$ are in $L^\infty([0,t] \times A \times R^1)$. Moreover, assume that there is a process $(\gamma_s(x)) \in L^\infty([0,t] \times R^1, \widetilde{\mathbb{D}^{1,\infty}})$ such that $\gamma_s(\omega,.) \in C^1(R^1)$ a.e., $(\frac{\partial}{\partial x}\gamma_s(x)) \in L^\infty([0,t] \times \Omega \times R^1)$ and $(\gamma_s(x)) = (\frac{\partial}{\partial x}a_s(x))$ a.e. on A.*

Set $\varphi(t) = \int\limits_0^t \|\gamma_s\|_{L^\infty(\Omega \times R^1)}ds$. Then, for any $X_o \in L^1(A)$, there is at most one solution $(X_s) \in \mathcal{D}_t(A)$ of the Skorohod stochastic differential equation

$$(3.2.12) \qquad dX_s = a_s(X_s)dW_s + b_s(X_s)ds, \qquad 0 \le s \le t,$$

on $B_{r-3\varphi(t)}(h)$.

Proof: Assume that we are given two solutions (X_s) and (Y_s) of equation (3.2.12) which belong to $\mathcal{D}_t(A)$. For all $0 \le s \le t$, we set

$$\tilde{\sigma}_s = \int\limits_0^1 (\partial_x a_s)(\theta X_s + (1-\theta)Y_s)d\theta,$$

$$\tilde{b}_s = \int\limits_0^1 (\partial_x b_s)(\theta X_s + (1-\theta)Y_s)d\theta.$$

Clearly, $(\tilde{\sigma}_s)$ and $(\tilde{b}_s)$ are in $L^\infty([0,t] \times A)$. From the assumption $(X_s),(Y_s) \in \mathcal{D}_t(A)$ we know that there are extensions $(\tilde{X}_s)$, $(\tilde{Y}_s)$ of (X_s) and (Y_s), respectively, with $(D_u\tilde{X}_s),(D_u\tilde{Y}_s) \in L^\infty([0,1] \times [0,t] \times \Omega)$. Hence, $(\sigma_s = \int\limits_0^1 \gamma_s(\vartheta\tilde{X}_s + (1-\vartheta)\tilde{Y}_s)d\vartheta)$ is an extension of $(\tilde{\sigma}_s)$ and belongs to $L^\infty([0,t] \times \Omega)$, its derivative

$$D_u\sigma_s = \int\limits_0^1 (D_u\gamma_s)(\vartheta\tilde{X}_s + (1-\vartheta))\tilde{Y}_s)d\vartheta + \int\limits_0^1 (\partial_x\gamma_s)(\vartheta\tilde{X}_s + (1-\vartheta)\tilde{Y}_s)\times$$
$$\times(\vartheta D_u\tilde{X}_s + (1-\vartheta)D_u\tilde{Y}_s)d\vartheta$$

exists and is in $L^\infty([0,1] \times [0,t] \times \Omega)$. Therefore, $(\sigma_s) \in L^\infty([0,t], \mathbb{D}^{1,\infty})$. Set $Z_s = X_s - Y_s$, $0 \le s \le t$. Since $(X_s),(Y_s) \in L^1([0,t] \times A)$ are solutions of (3.2.12), we have

$$(Z_s) \in L^1([0,t] \times A),$$
$$(\tilde{\sigma}_u Z_u I_{[0,s]}(u) = a_u(X_u)I_{[0,s]}(u) - a_u(Y_u)I_{[0,s]}(u)) \in \mathrm{Dom}\,\delta_A, \quad 0 \le s \le t,$$

and

$$Z_s = \int_0^s (a_u(X_u) - a_u(Y_u))dW_u + \int_0^s (b_u(X_u) - b_u(Y_u))du$$

$$= \int_0^s \tilde{\sigma}_u Z_u dW_u + \int_0^s \tilde{b}_u Z_u du \quad \text{a.e. on } A,\ 0 \le s \le t.$$

Consequently, all assumptions of Theorem 3.2.2 are satisfied such that $(Z_s) = 0$ is the unique solution of this linear equation on $B_{r-3\tilde{\varphi}(t)}(h)$ with

$$\tilde{\varphi}(t) = \int_0^t \|\sigma_s\|_\infty ds \le \int_0^t \|\gamma_s\|_{L^\infty(\Omega \times R^1)} ds.$$

This completes the proof.

Finally, let us consider linear stochastic differential equations with a drift function (σ_s) that is not smooth in the sense of the Malliavin calculus, but which is adapted to some enlargement of the filtration of the canonical process (W_s). More precisely, we suppose that (σ_s) is of the form

(3.2.13) $$\sigma_s(\omega) = K_s(\omega, G(\omega)), \quad (s, \omega) \in [0, 1] \times \Omega,$$

where $G \in \widetilde{I\!D^{1,\infty}}$ and, with the notations $a = \|G\|_\infty$ and $R_a = [-a, a]$, the function $K.(.,.) : [0, 1] \times \Omega \times R_a \to R^1$ is measurable and satisfies the following assumptions (K):

(i) $K.(., x)$ is adapted to the filtration generated by the canonical process, for each $x \in R^1$.

(ii) The funtion $K_s(\omega, .)$ belongs to $C_b^2(R^1)$ for a.e. $(s, \omega) \in [0, 1] \times \Omega$,

(iii) The norm

$$\|K\|_{(1)} = \|(\int_0^1 K_r^2 dr)^{1/2}\|_{L^\infty(\Omega \times R_a)} + (\int_0^1 \|K_r'\|_{L^\infty(\Omega \times R_a)}^2 dr)^{1/2} +$$

$$+ \|(\int_0^1 K_r''^2 dr)^{1/2}\|_{L^4(\Omega \times R_a)}$$

is finite.

For such a drift function (σ_s) we can state the following:

Theorem 3.2.7 *Let (σ_s) be of the form (3.2.13) with $G \in \widetilde{I\!D^{1,\infty}}$, and suppose that $(K_s(\omega, x))$ satisfies (K). Denote by $\{A_{s,t} : 0 \le s \le t \le 1\}$ the unique family of* *absolutely continuous transformations* $A_{s,t} : \Omega \to \Omega$ *associated to* (σ_s) *by Theorem 2.3.1 via the equation*

$$(3.2.14) \qquad A_{s,t}\omega = \omega - \int_{s\wedge\cdot}^{t\wedge\cdot} \sigma_r(A_{r,t}\omega)dr, \quad a.e., \quad 0 \le s \le t \le 1.$$

Further, denote the density presented in (2.3.3) by $L_{s,t}$, *and set* $A_t = A_{o,t}$ *and* $L_t = L_{o,t}$, $0 \le t \le 1$. *Then, for any random variable* η *in some* $L^q(\Omega)$, $q > 2$, *and any* $(b_s) \in L^1([0,1], L^\infty(\Omega))$, *the process*

$$(3.2.15) \qquad X_t = \eta(A_t) \exp\{\int_0^t b_s(A_{s,t})ds\} L_t, \quad 0 \le t \le 1,$$

is a global solution of equation (3.2.1). Moreover, $(X_t) \in L^2([0,1] \times \Omega)$ *is such that* $X_t \in L^2(\Omega)$, $0 \le t \le 1$, *and for any* $0 \le r \le 1$, *the process* $(X_t(A_{t,r})I_{[0,r]}(t))$ *is pathwise continuous in* $[0,r]$ *and bounded in* $L^2(\Omega)$.

Proof: Use the approximating sequence $(\sigma_s^n(\omega) = K_s^n(\omega, G^n(\omega)))$ for $(\sigma_s(\omega) = K_s(\omega, G(\omega))$ introduced in connection with Lemma 2.3.3, denote by $A_{s,t}^n : \Omega \to \Omega$, $0 \le s \le t \le 1$, the transformations associated to (σ_s^n) by (3.2.14) and define $L_{s,t}^n$ by (2.3.3) for (σ_s^n). For convenience we set $A_t = A_{o,t}$, $L_t = L_{o,t}$ $(0 \le t \le 1)$. As proved for Theorem 2.3.1, the sequence $(\sigma_s^n(A_{s,t}^n))$ converges in $L^2([0,t] \times \Omega)$ to $(\sigma_s(A_{s,t}))$ (Remark to Lemma 2.3.6), $L_{s,t}^n$ converges in $L^p(\Omega)$ to $L_{s,t}$ for all $0 \le s \le t \le 1$, and $\{L_{s,t}^n, (L_{s,t}^n)^{-1}, 0 \le s \le t \le 1, n = 1,2,3,\ldots\}$ is bounded in $L^p(\Omega)$ for all p $(1 < p < \infty)$. Hence, the Remark to Proposition 2.1.8 allows us to deduce that the random variables $\eta(A_t^n)$ and $b_s(A_{s,t}^n)$ converge in probability to $\eta(A_t)$ and $b_s(A_{s,t})$, respectively, so that the requirement $(b_s) \in L^1([0,1], L^\infty(\Omega))$ provides

$$\lim_{n\to\infty} \exp\{\int_0^t b_s(A_{s,t}^n)ds\} = \exp\{\int_0^t b_s(A_{s,t})ds\}, \quad 0 \le t \le 1,$$

in $L^p(\Omega)$, $1 < p < \infty$.
Moreover,

$$\begin{aligned} E[|\eta(A_t^n)L_t^n|^{q'}] &\le E[|\eta(A_t)|^q L_t^n]^{q'/q} E[(L_t^n)^{qq'/(q-q')}]^{(q-q')/q} \\ &\le E[|\eta|^q]^{2/q} \sup_{t,n} E[(L_t^n)^{qq'/(q-q')}]^{(q-q')/q}, \end{aligned}$$

for any q', $2 < q' < q$, $0 \le t \le 1$, $n = 1,2,3,\ldots$, implies

$$L^{q'}(\Omega) - \lim_{n\to\infty} \eta(A_t^n)L_t^n = \eta(A_t)L_t, \quad 0 \le t \le 1, \quad \text{and}$$

$$L^{q'}([0,1] \times \Omega) - \lim_{n\to\infty} (\eta(A_t^n)L_t^n) = (\eta(A_t)L_t), \quad \text{for all } q' \text{ s.t. } 2 < q' < q.$$

Therefore, with the notation

$$X_t^n = \eta(A_t^n)L_t^n \exp\{\int_0^t b_s(A_{s,t}^n)ds\},$$

we have for all q' $(2 < q' < q)$:

(3.2.16)
$$X_t = L^{q'}(\Omega) - \lim_{n\to\infty} X_t^n, \quad 0 \le t \le 1, \quad \text{and}$$
$$(X_t) = L^{q'}([0,1] \times \Omega) - \lim_{n\to\infty}(X_t^n).$$

Since $(\sigma_s^n) \in \mathbb{L}^S$ by construction, we know from Theorem 3.2.1 that the process (X_t^n) is a solution of the equation

$$X_t^n = \eta + \int_0^t \sigma_s^n X_s^n dW_s + \int_0^t b_s X_s^n ds, \quad 0 \le t \le 1,$$

i.e., for any $F \in \mathcal{S}$ it holds

(3.2.17) $\quad E[X_t^n F] = E[\eta F] + E[\int_0^t \sigma_s^n X_s^n D_s F ds] + E[\int_0^t b_s X_s^n ds \cdot F], \ 0 \le t \le 1.$

Due to relation (3.2.16) and the mean square convergence of the process (σ_s^n) to (σ_s) we can pass to the limit in (3.2.17), so that $(X_t) \in L^2([0,1] \times \Omega)$ is a solution of equation (3.2.1), indeed. In order to complete the proof note that the semi-group property of the family $(A_{s,t})$ of transformations, $A_{s,t} \circ A_{t,r} = A_{s,r}$, $0 \le s \le t \le r \le 1$, allows us to derive from the formulas (2.3.3) and (2.3.13) that

$$X_t(A_{t,r}) = \eta(A_r) \exp\{\int_0^t K_s(G(A_{s,r}))\bar{d}W_s - \frac{1}{2}\int_0^t K_s(G(A_{s,r})^2 ds$$
$$- \int_0^t K_s'(G(A_{s,r}))(D_s G)(A_{s,r})ds + \int_0^t b_s(A_{s,r})ds\}, \quad 0 \le t \le r \le 1,$$

where

$$\int_0^t K_s(G(A_{s,r}))\bar{d}W_s = \int_0^t K_s(x)dW_{s/x=G} -$$
$$- \int_0^r (\int_0^{u \wedge t} K_s'(x)dW_s)_{/x=G(A_{u,r})}) \times \frac{d}{du}[G(A_{u,r})]du.$$

So it is obvious that $t \mapsto X_t(A_{t,r})$ is continuous in $[0,r]$, and a straightforward computation gives

$$E[|X_t(A_{t,r})|^2] \le E[L_{t,r}|X_t(A_{t,r})|^{2+\delta}]^{2/2+\delta} \cdot E[(L_t^{-1})^{2+\delta/\delta}]^{\delta/2+\delta}$$
$$\le \sup_t E[|X_t|^{2+\delta}] \cdot \sup_{t \le r,n} E[(L_{t,r}^n)^{-2+\delta/\delta}]^{\delta/2+\delta}.$$

Note that the right-hand side is finite if $0 < \delta < q' - 2$. This completes the proof.

Since the transformations $A_{s,t}$, $0 \le s \le t \le 1$, associated with (σ_s) by (3.2.14) are not necessarily invertible, we should expect that the statement about the uniqueness of the solution of equation (3.2.1) is weaker than that of Theorem 3.2.1.

Theorem 3.2.8 *Suppose that the assumptions of Theorem 3.2.7 are satisfied, and, moreover, that there exists a process $(\hat{b}_t) \in L^\infty([0,1] \times \Omega)$ such that $\hat{b}_t(A_t) = b_t$, a.e., $0 \le t \le 1$. Then, for any solution $(X_t) \in L^2([0,1] \times \Omega)$ of equation (3.2.1) satisfying $X_t \in L^2(\Omega)$, $0 \le t \le 1$, and $\{X_t(A_{t,r}), 0 \le t \le r\}$ is continuous and bounded in $L^2(\Omega)$ for all $0 \le r \le 1$, it holds*

$$(3.2.18) \qquad E[X_t|\mathcal{A}_t] = \eta(A_t) \exp\{\int_0^t \hat{b}_s(A_t)ds\} E[L_t|\mathcal{A}_t], \quad a.e. \ 0 \le t \le 1.$$

Here $\mathcal{A}_t$ denotes the σ-field generated by the process $((A_t\omega)_s = \omega_s - \int_0^{s \wedge t} \sigma_r(A_{r,t}\omega)dr)$.

Proof: We use the notations introduced in Theorem 3.2.7. Since $(X_s) \in L^2([0,1] \times \Omega)$ is a global solution and, for any $F \in \mathcal{S}$, the random variable $F(A_t^n)$ is in $\mathbb{D}^{1,\infty}$, we obtain the following global variant in analogy to the computation (3.2.8).

$$(3.2.19) \qquad\qquad E[X_t F(A_t^n)] - E[\eta F] - E[\int_0^t b_s X_s F(A_s^n)ds] =$$

$$= E[\int_0^t (\sigma_s - \sigma_s^n)X_s D_s[F(A_s^n)]ds].$$

By the assumptions of the theorem and the bounded convergence of $F(A_t^n)$ to $F(A_t)$ in probability (cf. the proof of $\eta(A_t^n) \to \eta(A_t)$ for Theorem 3.2.7) we can pass to the limit on the left-hand side of (3.2.19). We can do the same on the right-hand side, too, since $\{D_s[F(A_s^n)], 0 \le s \le 1, n = 1, 2, 3, \ldots\}$ is essentially bounded. This can be easily derived from the equations

$$D_t[G^n(A_{s,t}^n)] = (D_t G^n)(A_{s,t}^n) - \int_s^t (D_v G^n)(A_{s,t}^n)(\partial_x K_v^n)(A_{v,t}^n)D_t[G^n(A_{v,t}^n)]dv,$$

and

$$D_t[F(A_t^n)] = (D_t F)(A_t^n) - \int_0^t (D_v F)(A_t^n)(\partial_x K_v^n)(A_{v,t}^n)D_t[G^n(A_{v,t}^n)]dv, \ 0 \le t \le 1,$$

by taking into account the assumptions on the approximating sequence $(K_r^n(\omega, G^n(\omega)))$ (cf. proof of Theorem 2.2.8).
After passing to the limit in (3.2.19) we have

$$(3.2.20) \qquad\qquad E[X_t F(A_t)] = E[\eta F] + E[\int_0^t b_s X_s F(A_s)ds].$$

Since $b_s = \hat{b}_s(A_s)$, the semigroup property of $(A_{s,t})$ provides $b_s(A_{s,r}) = \hat{b}_s(A_s)$, $0 \le s \le r \le 1$, and so we obtain by Girsanov transformation in (3.2.20):

$$E[X_t(A_{t,r})L_{t,r}F(A_r)] \quad = \quad E[\eta(A_r)L_rF(A_r)] + E[\int_0^t \hat{b}_s(A_r)X_s(A_{s,r})L_{s,r}ds \cdot F(A_r)]$$

for all $F \in \mathcal{S}$, i.e.,

$$E[X_t(A_{t,r})L_{t,r}|\mathcal{A}_r] = \eta(A_r)E[L_r|\mathcal{A}_r] + \int_0^t \hat{b}_s(A_t)E[X_s(A_{s,r})L_{s,r}|\mathcal{A}_r]ds,$$

a.e., $0 \leq t \leq r \leq 1$.
Hence,

$$E[X_t(A_{t,r})L_{t,r}|\mathcal{A}_r] = \eta(A_r)E[L_r|\mathcal{A}_r]\exp\{\int_0^t \hat{b}_s(A_r)ds\}, \text{ a.e., } 0 \leq t \leq 1.$$

On account of the assumptions on the process $\{X_t(A_{t,r}), 0 \leq t \leq r\}$, the bounded-ness of $\{L_{t,r}, 0 \leq t \leq r\}$ in $L^p(\Omega)$ for any $p > 1$ (recall Lemma 2.3.4 and $L_{t,r} = L^p(\Omega) - \lim_{n \to \infty} L_{t,r}^n$) and the continuity of $t \mapsto L_{t,r}$, $t \in [0, r]$ (cf. (2.3.13)) we can take the limit $t \to r$. This provides the desired relation (3.2.18).

Remark: In order to compute another expression for (3.2.18) recall that the transfor-mation $A_t : \Omega \to \Omega$ and the strictly positive density L_t satisfy the relation

$$E[F(A_t)L_t] = E[F], \quad F \in L^\infty(\Omega),$$

i.e., $P \circ [A_t]^{-1}$ is equivalent to P, the density $\mathcal{L}_t = \dfrac{dP \circ [A_t]^{-1}}{dP}$ is strictly positive, and

$$E[L_t|\mathcal{A}_t] = \mathcal{L}_t^{-1}(A_t), \quad \text{a.e.}$$

Obviously, (3.2.18) can now be expressed as follows:

(3.2.21) $$E[X_t|\mathcal{A}_t] = (\eta\exp\{\int_0^t \hat{b}_s ds\}\mathcal{L}_t^{-1} \circ A_t, \quad \text{a.e., } 0 \leq t \leq 1.$$

Under the assumption of a vanishing process (b_s) (3.2.21) provides

(3.2.22) $$E[X_t|\mathcal{A}_t] = (\eta\mathcal{L}_t^{-1}) \circ A_t, \quad \text{a.e., } 0 \leq t \leq 1.$$

If, moreover, (σ_t) is adapted to the filtration $(\mathcal{F}_t)$ of the canonical process (W_t) and η is deterministic, then the uniquely determined $(\mathcal{F}_t)$-adapted solution (X_t) is of the form

(3.2.23) $$X_t = \eta L_t, \quad \text{a.e., } 0 \leq t \leq 1.$$

Clearly, statement (3.2.22) is much weaker than (3.2.23), although we have (3.2.23) for the case of $(\mathcal{F}_t)$-adaptedness only. On the other hand, if $A_t = \mathcal{F}_1$ $(0 \leq t \leq 1)$, or equivalently, if the equation

$$\mathcal{E}_s^{(t)}(\omega) = \omega_s + \int_0^{s \wedge t} \sigma_r(\mathcal{E}^{(r)}(\omega))dr, \quad 0 \leq s, t \leq 1,$$

has a strong solution $\mathcal{E} = \{\mathcal{E}_s^{(t)}(\omega), 0 \leq s, t \leq 1\}$, then (3.2.18) implies that there is only one solution (X_t) that is given by (3.2.15).

3.3   Nonlinear Skorohod stochastic differential equations

Throughout this section we suppose that a and b are C^2-functions of R^1 into themselves whose derivatives of first and second order are bounded. This section mainly aims at giving a constructive proof of the existence of a solution of the Skorohod equation

$$(3.3.1) \qquad\qquad \begin{aligned} dX_t &= a(X_t)dW_t + b(X_t)dt, \qquad t \geq 0, \\ X_o &= \eta \end{aligned}$$

for certain initial values η that may depend on the whole path of the driving Wiener process (W_t).

Let us first introduce some notations and present the main results before coming to the rather long and technical details and proofs. For constructing the solution (X_s) of the equation (3.3.1) we need the associated Itô stochastic differential equation

$$(3.3.2) \quad X_t(x) = x + \int_0^t a(X_s(x))dW_s + \int_0^t b(X_s(x))ds, \quad 0 \leq s \leq 1, \quad x \in R^1.$$

On account of the pathwise description of the solution $(X_t(x))$, which was introduced and discussed by Doss and Sussmann, e.g. [43], we can define the transformation

$$(3.3.3) \qquad\qquad \begin{aligned} A_t(.,x) &: \quad \Omega \to \Omega, \\ A_t(\omega, x) &= \omega - \int_0^{t\wedge .} a'(X_s(\omega, x))ds, \quad \omega \in \Omega, \end{aligned}$$

for any fixed $(t, x) \in [0, 1] \times R^1$.

Let $\eta : \Omega \to R^1$ be any Fréchet differentiable bounded random variable satisfying the following assumptions (η):

$(\eta.1)$ For some real M it holds that
$|D_s\eta(\omega)| \leq M$, and
$|D_s\eta(\omega + \int_0^{\bullet} h_r dr) - D_s\eta(\omega)| \leq M|h|_2$, for all $h \in L^2([0,1])$, $(s, \omega) \in [0, 1] \times \Omega$.

$(\eta.2)$ There is a real $\delta > 0$ such that, with the notation $K = \|\eta\|_\infty$, we have

$$1 + \int_0^s (D_r\eta)(A_s(\omega, x))a''(X_r(\omega, x))\frac{\partial}{\partial x}X_r(\omega, x)dr \geq \delta,$$

for all $(s, \omega, x) \in [0, 1] \times \Omega \times [-K, K]$.

Then the Implicit Function Theorem provides:

Lemma 3.3.1 *Under the assumption (η) there is a unique process (U_s) with values in $[-K, K]$ that satisfies the equation*

$$(3.3.4) \qquad\qquad U_s(\omega) = \eta(A_s(\omega, U_s(\omega))), \quad (s, \omega) \in [0, 1] \times \Omega.$$

In particular, $U_o(\omega) = \eta(\omega)$. The process $(U_s(\omega))$ is pathwise absolutely continuous with respect to the Lebesgue measure, and for any $(s, \omega) \in [0, 1] \times \Omega$, there is a $(D_r U_s(\omega)) \in L^2([0, 1])$ such that, for all $h \in L^2([0, 1])$,

$$(3.3.5) \qquad \int_0^1 D_r U_s(\omega) h_r \, dr = \lim_{\varepsilon \downarrow 0} \frac{1}{\varepsilon} [U_s(\omega + \varepsilon \int_0^{\bullet} h_r \, dr) - U_s(\omega)].$$

This is the main argument for deriving of the following results:

Theorem 3.3.2 *Let a and b be $C^2(R^1)$-functions with bounded derivatives of first and second order. Then, under the condition (η), the process*

$$X_s(\omega) = X_s(\omega, U_s(\omega)), \qquad 0 \le s \le 1,$$

belongs to $\bigcap_{p>1} L^p([0, 1], D^{1,p})$ and satisfies (3.3.1),

$$X_s = \eta + \int_0^s a(X_r) \, dW_r + \int_0^s b(X_r) \, dr, \quad a.e., \, 0 \le s \le 1.$$

Proposition 3.3.3 *Under the assumption (η) the process (X_s) introduced in Theorem 3.3.2 has the following pathwise properties:*
For any $\omega \in \Omega$, the mapping $s \mapsto X_s(\omega)$ is continuous, and there is a real constant C such that, with the notation

$$\mathcal{E}(\omega) = \int_0^1 \exp\{C|\omega_s|\} \, ds,$$

the random variables $X_s(\omega)$ and $D_r X_s(\omega)$ are bounded by

$$\exp\{C(e^{C\mathcal{E}(\omega)} + |\omega_s|)\}, \quad \text{for all } 0 \le s, r \le 1, \, \omega \in \Omega.$$

In particular, for any bounded ball A in Ω, the process (X_s) belongs to $\mathcal{D}_1(A)$ (defined in preparation of Theorem 3.2.6).

From Proposition 3.3.3 we can derive the following statement about uniqueness:

Theorem 3.3.4 *Under the condition (η) the process (X_s) defined in Theorem 3.3.2 is the unique solution of the equation (3.3.1) inside the class of all processes that belong to $\mathcal{D}_1(A)$ for any bounded ball $A \subset \Omega$.*

Since the assumption $(\eta.2)$ is very restrictive, we should consider what happens if we omit $(\eta.2)$ and impose only $(\eta.1)$ on η. This will lead us to a local solution of (3.3.1):

Theorem 3.3.5 *Let $a, b \in C^2(R^1)$ such that the derivatives of first and second order are bounded, and assume $(\eta.1)$. Then, for any bounded ball $A = B_r(h)$, there is a $t > 0$ such that the Skorohod stochastic differential equation (3.3.1) has a unique solution $(X_s) \in \mathcal{D}_t(A)$ on $[0, t] \times B_{r-3\varphi(t)}(h)$, where $\varphi(t) = t \cdot \sup_x |a'(x)|$.*

After this presentation of the main results let us turn to the details now:

1. Review on the description of the solution of (3.3.1)

For any $x \in R^1$ let the continuous function $f : R^2 \to R^1$ be the solution of the equation

$$(3.3.6) \qquad f(x,y) = x + \int_0^y a(f(x,z))dz, \qquad y \in R^1.$$

Moreover, set

$$
\begin{aligned}
\Phi(x,y) &= (\frac{\partial}{\partial x}f(x,y))^{-1} \cdot (b - \frac{1}{2}a \cdot a')(f(x,y)), \\
\Psi(x,y) &= (\frac{\partial}{\partial x}f(x,y))^{-1} \cdot (b + \frac{1}{2}a \cdot a')(f(x,y)), \quad (x,y) \in R^2.
\end{aligned}
$$

Obviously, there exists a real C_1 such that the functions $\Phi(x,y)$, $\partial_x\Phi(x,y)$, $\Psi(x,y)$ and $\partial_s\Psi(x,y)$ are bounded by $C_1(1+|x|)\exp\{C_1|y|\}$, and $\partial_y\Phi(x,y)$ as well as $\partial_y\Psi(x,y)$ can be estimated by $C_2(1+|x|^2)\exp\{C_2|y|\}$, for some real C_2. Hence, there are unique processes $(\varphi_s(x))$ and $(\psi_s(x))$ satisfying pathwise the equations

$$
\begin{aligned}
(3.3.7) \qquad \varphi_s(\omega,x) &= x + \int_0^s \Phi(\varphi_r(\omega,x),\omega_r)dr, \\
\psi_s(\omega,x) &= x + \int_0^s \Psi(\psi_r(\omega,x),\omega_r)dr, \quad s \in [0,1],
\end{aligned}
$$

for all $(\omega,x) \in \Omega \times R^1$.
Finally, we define

$$
\begin{aligned}
(3.3.8) \qquad X_s(\omega,x) &= f(\varphi_s(\omega,x),\omega_s), \\
Y_s(\omega,x) &= f(\psi_s(\omega,x),\omega_s), \quad (s,\omega,x) \in [0,1] \times \Omega \times R^1.
\end{aligned}
$$

The processes $(X_s(x))$ and $(Y_s(x))$ defined in this way satisfy the Itô stochastic differential equations

$$X_s(x) = x + \int_0^s a(X_r(x))dW_r + \int_0^s b(X_r(x))dr,$$

and

$$Y_s(x) = x + \int_0^s a(Y_r(x))dW_r + \int_0^s (b + \frac{1}{2}a \cdot a')(Y_r(x))dr, \quad \text{a.e.}, \ 0 \le s \le 1,$$

respectively, which can be checked immediately by the nonanticipating Itô formula. Obviously, $X_s(\omega,.)$ and $Y_s(\omega,.)$ belong to $C^1(R^1)$, for all $(s,\omega) \in [0,1] \times \Omega$, and $X_s(.,x)$ as well as $Y_s(.,x)$ are Fréchet differentiable for all (s,x) in $[0,1] \times R^1$.

Lemma 3.3.6 *The processes* $(X_s(\omega, x))$, $(Y_s(\omega, x))$ *and their derivatives* $(\frac{\partial}{\partial x} X_s(\omega, x))$, $(\frac{\partial}{\partial x} Y_s(\omega, x))$ *are continuous in* $[0, 1] \times \Omega \times R^1$. *Moreover, there is a positive real* C *such that, with the notation*

$$\mathcal{E}(\omega) = \int_0^1 \exp\{C|\omega_r|\} dr, \quad \omega \in \Omega,$$

these processes as well as their Fréchet derivatives $(D_r X_s(\omega, x))$ *and* $(D_r Y_s(\omega, x))$ *are bounded by*

$$\exp\{(1 + |x|)e^{\mathcal{E}(\omega)} + C|\omega_s|\}, \quad \text{for all} \quad 0 \leq r, s \leq 1, \; \omega \in \Omega, \; x \in R^1.$$

Proof: The continuity of $X_s(\omega, x)$, $Y_s(\omega, x)$ and their derivatives $\frac{\partial}{\partial x} X_s(\omega, x)$, $\frac{\partial}{\partial x} Y_s(\omega, x)$ follows immediately from (3.3.8). It remains to prove the estimates. For this recall that we have

$$|\Phi(x, y)| \leq (1 + |x|) \exp\{C_2 |y|\}$$
$$|\frac{\partial}{\partial x} \Phi(x, y)| \leq (1 + |x|) \exp\{C_2 |y|\}$$
$$|\frac{\partial}{\partial y} \Phi(x, y)| \leq (1 + |x|^2) \exp\{C_2 |y|\},$$

for some real $C_2 > 0$, so that we can derive from (3.3.7) and the relations

$$(3.3.9) \qquad \frac{\partial}{\partial x} \varphi_s(\omega, x) = \exp\{\int_0^s (\frac{\partial}{\partial x} \Phi)(\varphi_r(\omega, x), \omega_r) dr\}.$$

$$D_r \varphi_s(\omega, x) = \int_r^s (\frac{\partial}{\partial y} \Phi)(\varphi_v(\omega, x), \omega_v) \frac{\frac{\partial}{\partial x} \varphi_s(\omega, x)}{\frac{\partial}{\partial x} \varphi_v(\omega, x)} dv \cdot I_{\{r \leq s\}}$$

that the variables $|\varphi_s(\omega, x)|$, $|\frac{\partial}{\partial x} \varphi_s(\omega, x)|$ and $|D_r \varphi(\omega, x)|$ are less than

$$(3.3.10) \qquad \qquad \exp\{(1 + |x|)e^{C_3 \mathcal{E}(\omega)}\}.$$

for some real $C_3 > 0$. Thus, with regard to (3.3.8) and the following two relations

$$\frac{\partial}{\partial x} X_s(\omega, x) = (\frac{\partial}{\partial x} f)(\varphi_s(\omega, x), \omega_s) \frac{\partial}{\partial x} \varphi_s(\omega, x),$$
$$D_r X_s(\omega, x) = (\frac{\partial}{\partial x} f)(\varphi_s(\omega, x), \omega_s) D_r \varphi_s(\omega, x) + (\frac{\partial}{\partial y} f)(\varphi_s(\omega, x), \omega_s) \cdot I_{\{r \leq s\}}$$

we see that the estimations in the lemma are correct. The estimations of $Y_s(\omega, x)$, $\frac{\partial}{\partial x} Y_s(\omega, x)$ and $D_r Y_s(\omega, x)$ can be deduced analogously.

Remark: The proof of Lemma 3.3.6 makes obvious that also the processes $(\varphi_s(\omega, x))$, $(\psi_s(\omega, x))$ and their derivatives are continuous with respect to (s, ω, x). The Fréchet derivatives $(D_r\varphi_s(\omega, x))$ and $(D_r\psi_s(\omega, x))$ are continuous with respect to (s, ω, x), uniformly relative to $0 \leq r \leq 1$.

In addition to the ω-wise estimation of $X_s(\omega, x)$ and $Y_s(\omega, x)$ we also need an a.e. estimation by a random variable of $L^*(\Omega)$ $(= \bigcap_{p>1} L^p(\Omega))$.

Lemma 3.3.7 *There are a real $q > 0$ and a random variable $\zeta \in L^*(\Omega)$ such that a.e. the random variables*

$$X_s(x), \quad \frac{\partial}{\partial x}X_s(x), \quad D_rX_s(x) \quad and \quad Y_s(x)$$

are bounded by $\zeta(1 + |x|^q)$ for all $0 \leq s, r \leq 1$, $x \in R^1$. If, additionally, the function a has a bounded derivative of third order, then the same is true also for $\frac{\partial}{\partial x}Y_s(x)$ and $D_rY_s(x)$.

Proof: The Lemmata 2.1 and 2.2 of [38], including the proofs, provide the necessary estimations.

From Lemma 3.3.7 we deduce an a.e. estimate by a random variable of $L^*(\Omega)$ also for $\varphi_t(x)$, $\psi_t(x)$ and their derivatives.

Lemma 3.3.8 *There exist a real $q > 0$ and a random variable of $L^*(\Omega)$ which are such that a.e. the random variables*

$$\varphi_s(x), \quad \frac{\partial}{\partial x}\varphi_s(x), \quad D_r\varphi_s(x) \quad and \quad \psi_s(x)$$

are less than

$$\zeta(1 + |x|^q), \quad for\ all \quad 0 \leq r, s \leq 1,\ x \in R^1.$$

Moreover, if a has a bounded derivative of third order, then the same holds true for $\frac{\partial}{\partial x}\psi_s(x)$ and $D_r\psi_s(x)$.

Proof: Without loss of generality we prove the statement for $\varphi_s(x)$, $\frac{\partial}{\partial x}\varphi_s(x)$ and $D_r\varphi_s(x)$ only. We first estimate $\varphi_s(x)$. Note that, by virtue of (3.3.7), there are reals $C_1, C_2 > 0$ such that

$$
\begin{aligned}
|\varphi_s(x)| &\leq |x| + \int_0^s e^{C_1|\omega_r|}|(b - \frac{1}{2}aa')(X_r(x))|dr \\
&\leq |x| + C_2 \int_0^s e^{C_1|\omega_r|}(1 + |X_r(x)|)dr.
\end{aligned}
$$

Substituting now the estimate for $X_r(x)$ of Lemma 3.3.7 we see that the assertion for $\varphi_s(x)$ is true. The estimate for $\frac{\partial}{\partial x}\varphi_s(x)$ we obtain from the relation

$$\frac{\partial}{\partial x}\varphi_s(x) = (\frac{\partial}{\partial x}f)(\varphi_s(x), \omega_s)\frac{\partial}{\partial x}X_s(x)$$

and Lemma 3.3.7, the assertion for $D_r\varphi_s(x)$ can be proved by substituting the estimates of $\varphi_s(x)$ and $D_r X_s(\omega)$ in the relation

$$D_r\varphi_s(x) = (\frac{\partial}{\partial x}f)(\varphi_s(x), \omega_s)^{-1}(D_r X_s(x) - (\frac{\partial}{\partial y}f)(\varphi_s(x), \omega_s)I_{\{r \leq s\}}).$$

This completes the proof.

2. Transformations generated by $(X_s(x))$ and $(Y_s(x))$.

For any $x \in R^1$, $0 \leq s \leq 1$, we define the following transformations of Ω into itself:

$$A_s(\omega, x) = \omega - \int_0^{s\wedge \cdot} a'(X_r(\omega, x))dr,$$

$$T_s(\omega, x) = \omega + \int_0^{s\wedge \cdot} a'(Y_r(\omega, x))dr, \quad \omega \in \Omega.$$

Lemma 3.3.9 *For any $x \in R^1$, $0 \leq s \leq 1$, the transformations $A_s(x)$ and $T_s(x)$ are inverse to each other.*

Proof: Let $h \in C([0, 1])$. By virtue of (3.3.7) and (3.3.8), the differentiation of $Y_s(\int_0^\cdot h_r dr, x)$ and $X_s(\int_0^\cdot h_r dr, x)$ relative to s provides

$$Y_s(\int_0^\cdot h_r dr, x) = x + \int_0^s a(Y_v(\int_0^\cdot h_r dr, x))h_v dv + \int_0^s (b + \frac{1}{2}aa')(Y_v(\int_0^\cdot h_r dr, x))dv,$$

$$X_s(\int_0^\cdot h_r dr, x) = x + \int_0^s a(X_v(\int_0^\cdot h_r dr, x))h_v dv + \int_0^s (b - \frac{1}{2}aa')(X_v(\int_0^\cdot h_r dr, x))dv,$$

$$0 \leq s \leq 1.$$

Substituting $T_1(\int_0^\cdot h_r dr, x) \in C_o^1([0, 1])$ for $\int_0^\cdot h_r dr$ in the second equation shows $(X_s(T_1(\int_0^\cdot h_r dr, x), x))$ to be a solution of the first equation. Thus, the uniqueness of the solution of these differential equations implies that $Y_s(\int_0^\cdot h_r dr, x)$ and $X_s(T_1(\int_0^\cdot h_r dr, x), x)$ coincide.

In order to conclude the equality of $Y_s(\omega, x)$ and $X_s(T_1(\omega, x), x)$ for all $\omega \in \Omega$, we only have to recall that the functions $\omega \mapsto Y_s(\omega, x)$ and $\omega \mapsto X_s(\omega, x)$ are continuous, since this also implies the continuity of $\omega \mapsto X_s(T_1(\omega, x), x)$ $(= X_s(\omega + \int_0^\cdot a'(Y_r(\omega, x))dr, x))$.

Therefore, $Y_s(\omega, x) = X_s(T_1(\omega, x), x)$, $(s, \omega, x) \in [0, 1] \times \Omega \times R^1$.

Since $(X_s(\omega, x))$ is nonanticipating, $X_s(T_1(\omega, x), x)$ and $X_s(T_t(\omega, x), x)$ coincide for all

$s \leq t$. Hence, for any $0 \leq s \leq 1$, we have

$$
\begin{aligned}
A_s(T_s(\omega, x), x) &= T_s(\omega, x) - \int_0^{s \wedge \cdot} a'(X_r(T_s(\omega, x), x)) dr \\
&= T_s(\omega, x) - \int_0^{s \wedge \cdot} a'(Y_r(\omega, x)) dr = \omega, \quad \omega \in \Omega.
\end{aligned}
$$

Analogously, for any $0 \leq s \leq 1$, $x \in R^1$, we obtain

$$
T_s(A_s(\omega, x), x) = \omega, \quad \omega \in \Omega.
$$

This completes the proof.

3. The process (U_s).

Let $\eta : \Omega \to R^1$ be any Fréchet differentiable bounded random variable satisfying the assumtions (η). We consider the equation $v_\bullet = \eta(A_\bullet(\bullet, v_\bullet))$. Its solution provides the process (U_s), which we have to substitute in $(X_s(x))$ in order to obtain a solution of $(3.3.1)$. Hence, recall the Implicit Function Theorem, which we need in the following version:

Proposition 3.3.10 *Let K be a positive real. Assume that $f : [0,1] \times [-K, K] \to [-K, K]$ is a continuous function with bounded derivatives $\frac{\partial}{\partial s} f(s, x)$ and $\frac{\partial}{\partial x} f(s, x)$ for which*

(i) the mapping $x \mapsto \frac{\partial}{\partial s} f(s, x)$ is continuous, uniformly relative to $s \in [0, 1]$.

(ii) $(s, x) \mapsto \frac{\partial}{\partial x} f(s, x)$ is continuous, and there is a real $\delta > 0$ such that

$$
1 - \frac{\partial}{\partial x} f(s, x) \geq \delta, \quad \text{for all} \quad (s, x) \in [0, 1] \times [-K, K].
$$

Then, for any $0 \leq s \leq 1$, there is a unique solution $x = x_s \in [-K, K]$ of the equation $x = f(s, x)$, and the function $s \mapsto x_s$ is absolutely continuous,

$$
\frac{d}{ds} x_s = \frac{\frac{\partial}{\partial s} f(s, x)}{1 - \frac{\partial}{\partial x} f(s, x)}.
$$

By means of this proposition we can prove Lemma 3.3.1. We divide the proof into two parts and extend the statement.

Proof of Lemma 3.3.1:
Step 1: Existence and uniqueness of the process (U_s) and its pathwise absolute continuity with derivative

$$
\frac{d}{ds} U_s(\omega) = -\frac{a'(X_s(\omega, U_s(\omega)))(D_s \eta)(A_s(\omega, U_s(\omega)))}{1 + \int_0^s (D_r \eta)(A_s(\omega, U_s(\omega)) a''(X_r(\omega, U_s(\omega)))(\frac{\partial}{\partial x} X_r)(\omega, U_s(\omega)) dr}.
$$

For any fixed $\omega \in \Omega$ we set
$$f(s, x) = \eta(A_s(\omega, x)).$$
Then, under (η), the chain rule shows that $\frac{\partial}{\partial x} f(s, x)$ exists and has the form

$$\frac{\partial}{\partial x} f(s, x) = - \int_0^s (D_r \eta)(A_s(\omega, x)) a''(X_r(\omega, x)) \frac{\partial}{\partial x} X_r(\omega, x) dr.$$

Consequently, $\frac{\partial}{\partial x} f(s, x)$ satisfies the assumptions required in Proposition 3.3.10. For the proof of the existence of $\frac{\partial}{\partial s} f(s, x)$ set

$$\kappa_\varepsilon(v) = 0, \quad \text{for } v \le 0, \quad \kappa_\varepsilon(v) = \frac{1}{\varepsilon} v, \quad \text{for } v \in [0, \varepsilon], \quad \text{and}$$
$$\kappa_\varepsilon(v) = 1, \quad \text{for } v \ge \varepsilon,$$

where ε is any real greater than 0. Then, for any $h \in L^2([0,1])$, the chain rule shows the existence of $\frac{d}{ds} \eta(\omega + \int_0^\bullet \kappa_\varepsilon(s - v) h_v dv)$,

$$\frac{d}{ds} \eta(\omega + \int_0^\bullet \kappa_\varepsilon(s - v) h_v dv) = \int_0^1 (D_r \eta)(\omega + \int_0^\bullet \kappa_\varepsilon(s - v) h_v dv) \partial_s \kappa(s - r) h_r dr$$

$$= \frac{1}{\varepsilon} \int_{s-\varepsilon}^s (D_r \eta)(\omega + \int_0^\bullet \kappa_\varepsilon(s - v) h_v dv) h_r dr.$$

Thus,

(3.3.11) $$\eta(\omega + \int_0^\bullet \kappa_\varepsilon(t - v) h_v dv) = \eta(\omega) +$$

$$+ \int_0^t \{ \frac{1}{\varepsilon} \int_r^{r+\varepsilon} (D_r \eta)(\omega + \int_0^\bullet \kappa_\varepsilon(s - v) h_v dv) ds \} h_r dr, \quad 0 \le t \le 1.$$

Under (η),

$$|(D_r \eta)(\omega + \int_0^\bullet \kappa_\varepsilon(s - v) h_v dv) - (D_r \eta)(\omega + \int_0^{r \wedge \bullet} h_v dv)|$$

$$\le M (\int_{r-\varepsilon}^{r+\varepsilon} h_v^2 dv)^{1/2}, \quad \text{for all } r \le s \le r + \varepsilon.$$

Hence, the right-hand side of (3.3.11) tends to

$$\eta(\omega) + \int_0^t (D_r \eta)(\omega + \int_0^{r \wedge \bullet} h_v dv) h_r dr, \quad \text{as } \varepsilon \to 0.$$

On the other hand, the convergence of the left-hand side of (3.3.11) to $\eta(\omega + \int_0^{t\wedge \cdot} h_v dv)$ is obvious. Consequently, for any $h \in L^2([0,1])$, the function $s \mapsto \eta(\omega + \int_0^{s\wedge \cdot} h_v dv)$ is absolutely continuous and

$$\frac{d}{ds}\eta(\omega + \int_0^{s\wedge \cdot} h_v dv) = h_s \cdot (D_s\eta)(\omega + \int_0^{s\wedge \cdot} h_v dv).$$

Setting now $h_v = -a'(X_v(\omega, x))$, we see that $\frac{\partial}{\partial s} f(s, x)$ exists and is given by

$$\frac{\partial}{\partial s} f(s, x) = -a'(X_s(\omega, x))D_s\eta)(A_s(\omega, x)).$$

Clearly, $\frac{\partial}{\partial s} f(s, x)$ is bounded, $a'(X_s(\omega, x))$ is continuous relative to (s, x) and, thus, from the estimate

$$|(D_s\eta)(A_s(\omega, x)) - (D_s\eta)(A_s(\omega, y))| \leq M(\int_0^1 |a'(X_r(\omega, x)) - a'(X_r(\omega, y))|^2 dr)^{1/2},$$

$$x, y \in [-K, K], \ (s, \omega) \in [0, 1] \times \Omega,$$

we conclude that $x \mapsto \frac{\partial}{\partial s} f(s, x)$ is continuous, uniformly with respect to $0 \leq s \leq 1$. Thus, Proposition 3.3.10 can be applied, which yields the desired result.

Remark: In order to abbreviate the notations we introduce the transformations

$$A_s\omega = A_s(\omega, U_s(\omega)), \quad T_s\omega = T_s(\omega, \eta(\omega)), \quad \omega \in \Omega.$$

Since $T_s(x)$ and $A_s(x)$ are inverse to each other for all $x \in R^1$, equation (3.3.4) implies the same for T_s, A_s. Moreover, due to Theorem 2.4.1 the image measures $P \circ [T_s]^{-1}$, $P \circ [A_s]^{-1}$ are equivalent to the Wiener measure P.

Step 2 (of the proof of Lemma 3.3.1): $(D_r U_s(\omega))$ exists and is given by

$$D_r U_s(\omega) = \frac{(D_s\eta)(A_s\omega) - \int_0^s (D_v\eta)(A_s\omega)a''(X_v(\omega, U_s(\omega)))(D_r X_v)(\omega, U_s(\omega))dv \cdot I_{\{r \leq s\}}}{1 + \int_0^s (D_r\eta)(A_s\omega)a''(X_v(\omega, U_s(\omega)))(\frac{\partial}{\partial x} X_v)(\omega, U_s(\omega))dv}.$$

Fix any $(s, \omega) \in [0, 1] \times \Omega$ and any $h \in L^2([0,1])$, and set

$$f(\varepsilon, x) = \eta(A_s(\theta_{\varepsilon h}\omega, x))$$
$$= \eta(\theta_{\varepsilon h}\omega - \int_0^{s\wedge \cdot} a'(X_v(\theta_{\varepsilon h}\omega, x)dv).$$

For convenience we drop ω, although the following calculation will be performed for any fixed $\omega \in \Omega$. Note that

$$\frac{\partial}{\partial \varepsilon}[X_r(\theta_{\varepsilon h}, x)] = \int_0^r (D_v X_r)(\theta_{\varepsilon h}, x))h_v dv,$$

so that the chain rule yields

$$(3.3.12) \quad \frac{\partial}{\partial \varepsilon} f(\varepsilon, x) = \int_0^1 (D_r \eta)(A_s(\theta_{\varepsilon h}, x)) \times$$

$$\times \quad \{h_r - I_{\{r \leq s\}} a''(X_r(\theta_{\varepsilon h}, x)) \int_0^r (D_v X_r)(\theta_{\varepsilon h}, x) h_v dv) dr.$$

Analogously to the first step we can derive

$$(3.3.13) \quad \frac{\partial}{\partial x} f(\varepsilon, x) = - \int_0^1 (D_r \eta)(A_s(\theta_{\varepsilon h}, x)) a''(X_r(\theta_{\varepsilon h}, x))(\frac{\partial}{\partial x} X_r)(\theta_{\varepsilon h}, x) dr.$$

Obviously, in view of (3.3.12) and (3.3.13) the assumptions of Proposition 3.3.10 are satisfied here, too. Thus, the unique solution $(U_s(\theta_{\varepsilon h}))$ of the equation

$$U_s(\theta_{\varepsilon h}) = \eta(A_s(\theta_{\varepsilon h}, U_s(\theta_{\varepsilon h}))) = f(\varepsilon, U_s(\theta_{\varepsilon h})), \ 0 \leq \varepsilon \leq 1,$$

is absolutely continuous with respect to ε, and

$$\frac{d}{d\varepsilon} U_s(\theta_{\varepsilon h}) = \int_0^1 h_r \left[\frac{(D_r \eta)(A_s) - \int_r^s (D_v \eta) a''(X_v(U_s))(D_r X_v)(U_s) dv I_{\{r \leq s\}}}{1 + \int_0^s (D_v \eta)(A_s) a''(X_v(U_s))(\frac{\partial}{\partial x} X_v)(U_s) dv} \right] \circ \theta_{\varepsilon h} dr.$$

Setting $\varepsilon = 0$ we obtain the desired result. This completes the proof of Lemma 3.3.1.

Lemma 3.3.11 *Assume (η). Then, for any $p > 1$, the process (U_s) belongs to $L^p([0,1], \mathbb{D}^{1,p}) \cap L^\infty([0,1] \times \Omega)$,*

(i) the mapping $s \mapsto (D_r U_s) \in L^p([0,1] \times \Omega)$ is continuous, and

(ii) $s \mapsto D_r U_s \in L^p(\Omega)$ is continuous in $[0,r]$, uniformly with respect to r.

Moreover, the process $(\frac{d}{ds} U_s)$ is bounded by some real constant, $(D_r U_s)$ is bounded by a random variable of $L^*(\Omega)$, and, additionally, there is some real $C > 0$ such that, with the notation

$$\mathcal{E}(\omega) = \int_0^1 \exp\{C|\omega_s|\} ds,$$

we have

$$(3.3.14) \quad |D_r U_s(\omega)| \leq \exp\{e^{C\mathcal{E}(\omega)}\} \text{ for all } 0 \leq s, r \leq 1, \ \omega \in \Omega.$$

Proof: Obviously, the process (U_s) as well as its derivative $(\frac{d}{ds}U_s)$ are bounded by some real constant. By virtue of the second step of the proof of Lemma 3.3.1 there exists a real $C > 0$ such that

$$|D_r U_s(\omega)| \le C(1 + \int_r^s |(D_r X_v)(\omega, U_s(\omega))|dv \cdot I_{\{r \le s\}}), \quad \omega \in \Omega.$$

Then, the Lemmata 3.3.6 and 3.3.7 imply that $(D_r U_s)$ is bounded by a random variable of $L^*(\Omega)$ and that there is a real C with (3.3.14).

Thus, it remains to show (i) and (ii). We first turn to (ii). Let $s \le r$. Then step 2 of the proof of Lemma 3.3.1 provides

$$(3.3.15) \qquad D_r U_s = \frac{(D_r \eta)(A_s)}{1 + \int_0^s (D_v \eta)(A_s) a''(X_v(U_s))(\frac{\partial}{\partial x}X_v)(U_s)dv}.$$

Taking into account that

$$A_s \omega = \omega - \int_0^{s \wedge \cdot} a'(X_v(\omega, U_s(\omega)))dv,$$

the Lipschitz condition $(\eta.1)$ yields for all $0 \le r \le 1$ and all $0 \le s \le t \le 1$ that

$$(3.3.16) \quad |(D_r \eta)(A_t) - (D_r \eta)(A_s)| \le M(\int_0^s |a'(X_v(U_t)) - a'(X_v(U_s))|^2 dv)^{1/2}$$

$$+ M(\int_s^t |a'(X_v(U_t))|^2 dv)^{1/2}.$$

Since

$$|a'(X_v(U_t)) - a'(X_v(U_s))| \le$$

$$\le \sup_x |a''(x)| \cdot \sup_r |\frac{d}{dr}U_r| \cdot \int_0^1 |(\frac{\partial}{\partial x}X_v)(\theta U_t + (1-\theta)U_s)|d\theta \cdot |t-s|,$$

we obtain

$$(3.3.17) \qquad |(D_r \eta)(A_t) - (D_r \eta)(A_s)| \le \zeta|t-s|, \quad \text{for all } 0 \le r, s, t \le 1.$$

from the estimate of $\frac{\partial}{\partial x}X_v(x)$ in Lemma 3.3.7 for some $\zeta \in L^*(\Omega)$. Consequently, for any $p > 1$, the mapping $s \mapsto (D_r \eta)(A_s) \in L^p(\Omega)$, $0 \le s \le r$, is continuous uniformly relative to $0 \le r \le 1$. Since, on the other hand, by virtue of $(\eta.2)$ and the Lemmata 3.3.1 and 3.3.6,

$$(3.3.18) \qquad s \mapsto 1 + \int_0^s (D_v \eta)(A_s) a''(X_v(U_s)(\frac{\partial}{\partial x}X_v)(U_s)dv \in L^p(\Omega)$$

is continuous and uniformly bounded, relation (3.3.15) now implies the correctness of statement (ii).

Finally, we prove (i). For this, note that by the same arguments as used above we can conclude that also the mapping

$$s \mapsto \int_r^s (D_v \eta)(A_s) a''(X_v(U_s))(D_r X_v)(U_s) dv \cdot I_{\{r \le t\}} \in L^p(\Omega)$$

is continuous for all $0 \le r \le 1$, and from the Lemmata 3.3.7 and 3.3.11 we can see

$$\sup_{0 \le r,s \le 1} | \int_r^s (D_v \eta)(A_s) a''(X_v(U_s))(D_r X_v)(U_s) dv \cdot I_{\{r \le s\}}| \in L^*(\Omega).$$

Together with (3.3.17) and (3.3.18) this gives (i).

By property (ii) of Lemma 3.3.11 we can define

$$(D_- U)_s = L^2(\Omega) - \lim_{r \to s, r > s} D_s U_r, \qquad 0 \le s \le 1.$$

An immediate consequence of the Lemmata 3.3.1 and 3.3.11 is given by the following statement:

Lemma 3.3.12 *Under the assumption (η), the process $((D_- U)_s)$ is bounded by some $\zeta \in L^*(\Omega)$, and*

$$\frac{d}{ds} U_s = -a'(X_s(U_s))(D_- U)_s, \qquad \text{for a.e. } 0 \le s \le 1.$$

4. Proof of the Theorems 3.3.2 and 3.3.4:

Note that by virtue of (3.3.8) the process (X_s) has the form

$$(3.3.19) \qquad\qquad X_s = f(V_s, W_s), \qquad 0 \le s \le 1,$$

where the process $V_s = \varphi_s(U_s)$ is absolutely continuous with respect to s. This is the reason why the Itô formula plays a key role in the proof of Theorem 3.3.2. Let us recall the anticipative Itô formula by Nualart and Pardoux (cf. Theorem 6.1, [33]) for the special case we need.

Proposition 3.3.13 *Let f be a continuous function of R^2 into R^1 such that the derivatives $\frac{\partial}{\partial x} f$, $\frac{\partial}{\partial y} f$, $\frac{\partial^2}{\partial y \partial x} f$ and $\frac{\partial^2}{\partial y^2} f$ exist and are continuous. Moreover, let (V_s) be a continuous process with finite variation belonging to $L^2([0,1], I\!\!D^{1,2})$ such that*

(i) $E[\int_0^1 \int_0^1 |D_r V_s|^4 dr ds] < +\infty$,

(ii) *the mapping $s \mapsto DV_s \subset L^4([0,1] \times \Omega)$ is continuous, and*

(iii) $s \mapsto D_r V_s \in L^4(\Omega)$ *is continuous in* $[0, r]$, *uniformly with respect to* $0 \leq r \leq 1$. *Then, for any* $0 \leq s \leq 1$, *with the notation*

$$(D_- V)_s = L^2(\Omega) - \lim_{r \to s, r > s} D_s V_r,$$

we have:

$$f(V_s, W_s) = f(V_o, 0) + \int_0^s \frac{\partial}{\partial x} f(V_r, W_r) dV_r + \int_0^s \frac{\partial}{\partial y} f(V_r, W_r) dW_r$$

$$+ \frac{1}{2} \int_0^s \frac{\partial^2}{\partial y^2} f(V_r, W_r) dr + \int_0^s \frac{\partial^2}{\partial y \partial x} f(V_r, W_r)(D_- V)_r dr.$$

If $(\frac{\partial}{\partial y} f(V_r, W_r) I_{[0,s]}(r)) \in Dom\, \delta$, *then* $\int_0^s \frac{\partial}{\partial y} f(V_r, W_r) dW_r$ *denotes the Skorohod integral, otherwise it is the local Skorohod integral.*

In Theorem 6.10 of [33] the Itô formula is established under the stronger assumption of the continuity of the process $s \mapsto D_r V_s \in L^4(\Omega)$, $0 \leq s \leq 1$, which is uniform with respect to $0 \leq r \leq 1$, but it turns out that the proof given in [33] needs the weaker assumptions (ii) and (iii) of Proposition 3.3.13 only.

Proof (of Theorem 3.3.2): We set that $V_s = \varphi_s(U_s)$. Then, obviously, the process (V_s) is absolutely continuous,

(3.3.20) $$\frac{d}{ds} V_s = \Phi(V_s, W_s) + (\frac{\partial}{\partial x} \varphi_s)(U_s) \frac{d}{ds} U_s, \quad 0 \leq s \leq 1,$$

and in view of

(3.3.21) $$D_r V_s = (D_r \varphi_s)(U_s) + (\frac{\partial}{\partial x} \varphi_s)(U_s) D_r U_s, \quad 0 \leq s, r \leq 1,$$

we can derive from the Remark to Lemma 3.3.6 and the Lemmata 3.3.8 and 3.3.11 that the process (V_s) satisfies the assumptions of Proposition 3.3.13. In particular, the process $((D_- V)_s)$ exists, and due to Lemma 3.3.12 we can deduce the relation

(3.3.22) $$\frac{d}{ds} V_s = \Phi(V_s, W_s) - a'(X_s)(D_- V)_s, \quad 0 \leq s \leq 1.$$

from (3.3.20) and (3.3.21).

Since also the solution $f : R^2 \to R^1$ of equation (3.3.6) has the required properties, we can apply the anticipative Itô formula to $X_s = f(V_s, W_s)$: Basing on the equations (3.3.6), (3.3.4) and (3.3.8) defining f, $X_s(x)$, U_s and X_s respectively we compute

$$f(V_o, 0) = \eta,$$

$$\frac{\partial}{\partial y} f(V_s, W_s) = a(X_s),$$

$$\frac{\partial^2}{\partial y^2} f(V_s, W_s) = (a \cdot a')(X_s),$$

and taking into consideration equation (3.3.22) we see that

$$
\begin{aligned}
\frac{\partial}{\partial x} f(V_s, W_s) \frac{d}{ds} V_s &= \frac{\partial}{\partial x} f(V_s, W_s)(\Phi(V_s, W_s) - a'(X_s)(D_- V)_s) \\
&= (b - \frac{1}{2} a \cdot a')(X_s) - \frac{\partial^2}{\partial y \partial x} f(V_s, W_s)(D_- V)_s.
\end{aligned}
$$

Hence, the anticipative Itô formula of Proposition 3.3.13 provides the equation (3.3.1). From the estimations of $X_s(x)$, $\frac{\partial}{\partial x} X_s(x)$ and $D_r X_s(x)$ as well as from U_s and $D_r U_s$ in Lemma 3.3.7 and Lemma 3.3.11, respectively, we see that the process (X_s) belongs to $L^p([0,1], I\!D^{1,p})$ for any $p > 1$. However, we obviously have $(a(X_r) I_{[0,s]}(r)) \in \mathrm{Dom}\, \delta$, for all $0 \leq s \leq 1$. Hence, (X_s) is a solution of the Skorohod equation (3.3.1).

Remark: Note that by virtue of

$$
X_s(\omega, x) = Y_s(A_s(\omega, x), x) \quad \text{and} \quad A_s \omega = A_s(\omega, U_s(\omega)),
$$

it also holds

$$
X_s(\omega) = Y_s(\eta) \circ A_s \omega, \qquad (s, \omega) \in [0,1] \times \Omega.
$$

Now we can prove Proposition 3.3.3 and then Theorem 3.3.4 by means of Theorem 3.3.2.

Proof (of Proposition 3.3.3): The proof of the estimate follows immediately from the estimates of $X_s(\omega, x)$, $\frac{\partial}{\partial x} X_s(\omega, x)$ and $D_r X_s(\omega, x)$ and those of $U_s(\omega)$ and $D_r U_s(\omega)$ in Lemma 3.3.6 and Lemma 3.3.11, respectively. Thus, for any bounded ball $A = B_r(h) \subset \Omega$ and any $H \in I\!D^{1,\infty}(B_{r+1}(h))$ with $\{H = 1\} \supset A$, the product (HX_s) is in $L^\infty([0,1], I\!D^{1,\infty})$, and, since (HX_s) coincides with (X_s) on A, (X_s) is in $\mathcal{D}_1(A)$.

Proof (of Theorem 3.3.4): Note that, for any bounded ball $A = B_r(h) \subset \Omega$, $r > 0$, $h \in L^2([0,1])$, the solution (X_s) of (3.3.1) on Ω is also a solution of (3.3.1) on A. Since, moreover, $(X_s) \in \mathcal{D}_1(A)$, we can apply Theorem 3.2.2 by setting $\gamma_s(\omega, x) = a'(x)$ and
$$
\varphi(t) = \int_0^t \|\gamma\|_{L^\infty(\Omega \times R^1)} ds = t \cdot \sup_x |a'(x)|.
$$
Hence, (X_s) is unique on $B_{r-3\varphi(t)}(h)$. This holds for any $r > 0$ and $h \in L^2([0,1])$, and so for Ω, too.

Remark: As we have shown, the Skorohod equation (3.3.1) has a unique solution under the assumption (η). The existence of such a solution on the whole probability space Ω is mainly ensured by the assumption $(\eta.2)$, while $(\eta.1)$ is only some smoothness requirement for η by means of which we can apply the machinery of the anticipative stochastic calculus. So we should pay some more attention to $(\eta.2)$. Obviously, we can replace $(\eta.2)$ by the much stronger condition $(\eta'.2)$, namely

$(\eta'.2)$ either

    (i) $D_s G(\omega) \geq 0$, for all $(s, \omega) \in [0,1] \times \Omega$, and $a''(x) \geq 0$, for all $x \in R^1$,
       or

(ii) $D_s G(\omega) \leq 0$, for all $(s, \omega) \in [0, 1] \times \Omega$, and $a''(x) \leq 0$, for all $x \in R^1$.

In particular, if $\eta = f(W_1)$, $f \in C^2(R^1)$, then the requirements (i) and (ii) take the form $f'(x) \geq 0$, $a''(x) \geq 0$, for all $x \in R^1$, and $f'(x) \leq 0$, $a''(x) \leq 0$, for all $x \in R^1$, respectively.

The price for omitting $(\eta.2)$ consists in the loss of the global existence of the solution of (3.3.1), the assumption $(\eta.1)$ guarantees only the existence of a local solution as stated by Theorem 3.3.5.

Proof (of Theorem 3.3.5): Let $\lambda \in C^\infty(R^1)$ with $\lambda(x) = |x|$ for $|x| \geq 1$, and $|x| \leq \lambda(x) \leq 1$, for $|x| \leq 1$. With regard to Lemma 3.3.6 there is some real $C > 0$ such that

$$\left(\int_0^1 |\frac{\partial}{\partial x} X_s(x)|^2 ds\right)^{1/2} \leq \exp\{C(1 + |x|)e^{\mathcal{E}(\omega)}\}, \quad (\omega, x) \in \Omega \times R^1,$$

for $\mathcal{E}(\omega) = \int_0^1 \exp\{C\lambda(\omega_s)\}ds$.

Fix any reals r_1, r_2 with $0 < r_1 < r_2$ and a $\kappa \in C_o^\infty(R^1)$ chosen such that

$$\kappa(x) = 1, \quad \text{for} \quad |x| \leq r_1, \quad \kappa(x) = 0, \text{ for } |x| \geq r_2, \text{ and}$$
$$0 \leq \kappa(x) \leq 1, \quad \text{for all} \quad x \in R^1,$$

and set $\tilde{\eta}(\omega) = \eta(\omega) \cdot \kappa(\mathcal{E}(\omega))$.

Due to the assumption, η satisfies $(\eta.1)$ and, obviously, so does $\tilde{\eta}$.
In particular,

$$\tilde{K} = \|\tilde{\eta}\|_\infty \leq \|\eta\|_\infty,$$
$$\|D\tilde{\eta}\|_{L^\infty([0,1]\times\Omega)} \leq \|D\eta\|_{L^\infty([0,1]\times\Omega)} + r_2 \cdot \sup_x |\kappa'(x)| \cdot \sup_x |\lambda'(x)| \cdot \|\eta\|_\infty,$$

and in virtue of the relation

$$\mathcal{E}(\omega) \leq \exp\{C(1 + \sup_x |a'(x)|)\}\mathcal{E}(A_s(\omega, x))$$

we have

$$\text{supp}\,(D_r\tilde{\eta})(A_s(x)) \subset \{|\mathcal{E}| \leq r_2'\} \quad \text{with} \quad r_2' = r_2 \exp\{C(1 + \sup_x |a'(x)|)\}.$$

Hence, for certain real $C(r_2)$ we have

$$\int_0^1 |(D_r\tilde{\eta})(A_s(\omega, x))a''(X_r(\omega, x))(\frac{\partial}{\partial x}X_r)(\omega, x)|dr \leq$$

$$\leq C(r_2)(\int_0^1 |(\frac{\partial}{\partial x}X_r)(\omega, x)|^2 dr)^{1/2}t^{1/2}I\{|\mathcal{E}| \leq r_2'\}$$

$$\leq C(r_2)\exp\{C(1 + \tilde{K})e^{r_2'}\}t^{1/2}$$
$$\text{for all} \quad (s, \omega, x) \in [0, t] \times \Omega \times [-\tilde{K}, \tilde{K}].$$

Let $0 < \delta < 1$. Then the expression in the last line is less than $1 - \delta$ if only $t = t(r_2) > 0$ is small enough. In this case we obtain

$$1 + \int_0^s (D_r \tilde{\eta})(A_s(\omega, x)) a''(X_r(\omega, x)) \partial_x X_r(\omega, x) dr \geq \delta,$$

$$\text{for all} \quad (s, \omega, x) \in [0, t] \times \Omega \times [-\tilde{K}, \tilde{K}],$$

i.e., $\tilde{\eta}$ satisfies also $(\eta.2)$ on the time interval $[0, t]$. Of course, this restriction of $(\eta.2)$ to $[0, t]$ does not matter if we consider the Skorohod stochastic differential equation (3.3.1) for this time interval only. Thus, Theorem 3.3.4 provides a solution $(\tilde{X}_s) \in \bigcap_{p>1} L^p([0, t], \mathbb{D}^{1,p})$ of the equation

$$\tilde{X}_s = \tilde{\eta} + \int_0^s a(\tilde{X}_r) dW_r + \int_0^s b(\tilde{X}_r) dr \quad \text{a.e.,} \ s \in [0, t],$$

which is unique in the class of all processes whose restriction to any bounded ball $A \subset \Omega$ belongs to $\mathcal{D}_t(A)$.

Since $\tilde{\eta} = \eta$ on $B_{r_1'}(0) \ (\subset \{|\mathcal{E}| \leq r_1\})$ for $r_1 = \frac{1}{C} \ln r_1 - 1$, $(\tilde{X}_s) \in \mathcal{D}_t(B_{r_1'}(0))$ is a solution of equation (3.3.1) on $B_{r_1'}(0)$ with initial value η. Hence, for a given ball $A = B_r(h) \subset \Omega$ choose $r_1' > r + |h|_{L^2([0,1])}$. Then $A \subset B_{r_1'}(0)$, and $(\tilde{X})$ is a solution of (3.3.1) on A and belongs to $\mathcal{D}_t(A)$. In order to complete the proof note that the uniqueness of the solution $(\tilde{X}_s) \in \mathcal{D}_t(A)$ on $B_{r-3t \sup_x |a'(x)|}(h)$ follows from Theorem 3.2.2.

6. Existence of a density of X_s

We assume again (η) and ask for a criterion for the absolute continuity of the solution $X_s \in \mathbb{D}^{1,*} \ (= \bigcap \mathbb{D}^{1,p})$ of the Skorohod stochastic differential equation (3.3.1) with respect to the Lebesgue measure for any $0 \leq s \leq 1$. Hence, we will use Theorem 7 of [19]:

Proposition 3.3.14 *If $F \in \mathbb{D}^{1,2}$ and $\int_0^1 |D_r F|^2 dr > 0$ a.e., then F has a density with respect to the Lebesgue measure.*

Now we can state the following:

Proposition 3.3.15 *Let a be a $C^3(R^1)$-function with bounded derivatives of the first three orders, b a $C^2(R^1)$-function with bounded derivatives of the first two orders, assume (η) and denote by (X_s) the solution of (3.3.1) presented in Theorem 3.3.2. Then, for any s, $0 < s \leq 1$, X_s has a density if*

$$(3.3.23) \qquad P(\{\int_s^1 |D_r \eta|^2 dr = 0\} \ \cap \ \{(D_r \eta) \in C^1([0, s]), D_o \eta = -a(\eta),$$

$$\frac{d}{ds} D_s \eta_{|s=0} \ = \ -[b + \frac{1}{2}aa', a](\eta)\}) = 0.$$

Here $[.,.]$ denotes the Lie brackets.

Proof: Since X_s is of the form

$$X_s = Y_s(\eta) \circ A_s, \qquad 0 \le s \le 1,$$

and $P \circ [A_s]^{-1}$ is equivalent to the Wiener measure (cf. Remark to the proof of Theorem 3.3.2), the random variable X_s has a density if and only if $Y_s(\eta)$ has a density. Consequently, we apply Proposition 3.3.14 to $Y_s(\eta)$. Note that $Y_s(\eta) \in \mathbb{D}^{1,*}$, and

$$D_r[Y_s(\eta)] = (\frac{\partial}{\partial x} Y_s)(\eta) D_r\eta + a(Y_r(\eta)) \frac{(\frac{\partial}{\partial x} Y_s)(\eta)}{(\frac{\partial}{\partial x} Y_r)(\eta)} I_{\{r \le s\}}, \qquad 0 \le r \le 1.$$

On the other hand, by the nonanticipative Itô formula we see that, for any $x \in R^1$,

$$(3.3.24) \quad a(Y_r(x))(\frac{\partial}{\partial x} Y_r(x))^{-1} = a(x) + \int_0^r [b + \frac{1}{2} aa', a](Y_v(x)) \frac{\partial}{\partial x} Y_v(x) dv,$$

$$\text{for all } 0 \le r \le 1, \text{a.e.}$$

Since both sides are continuous with respect to x, η can be substituted, which provides

$$D_r[Y_s(\eta)] = (\frac{\partial}{\partial x} Y_s)(\eta)\{D_r\eta + (a(\eta) + $$

$$+ \int_0^r [b + \frac{1}{2} aa', a](Y_v(\eta)) \cdot (\frac{\partial}{\partial x} Y_v)(\eta) dv) I_{\{r \le s\}}.$$

Hence,

$$\{\int_0^1 |D_r[Y_s(\eta)]|^2 dr = 0\} = \{\int_s^1 |D_r\eta|^2 dr = 0\} \cap$$

$$\cap \ \{\int_s^r |D_r\eta + a(\eta) + \int_0^r [b + \frac{1}{2} aa', a](Y_v(\eta)) \cdot (\frac{\partial}{\partial x} Y_v)(\eta) dv|^2 dr = 0\}.$$

On this set the mapping $r \mapsto D_r\eta$ has a modification which belongs to $C^1([0, s])$, takes the value $-a(\eta)$ for $r = 0$, and its derivative $\frac{d}{dr} D_r G$ has the value $-[b + \frac{1}{2} aa', a](\eta)$ in $r = 0$. This provides the desired result.

By the Taylor expansion of $a(Y_s(x))Y_s(x)^{-1}$, which was computed by S. Kusuoka and D. Stroock in [30], one can improve condition (3.3.23) by the following requirement: With probability zero, the Taylor expansions of $(D_r\eta)$ and $a(Y_s(x))Y_s(x)^{-1}$ coincide. Of course, this requires the more restrictive assumption that $a, b \in C^\infty(R^1)$ have bounded derivatives of all orders and the initial value η is a smooth random variable.

3.4 Skorohod stochastic differential equations with boundary condition

In this section we study the stochastic differential equation

$$(3.4.1) \qquad X_t = X_o + \int_0^t \sigma_s X_s dW_s + \int_0^t b_s X_s ds, \quad 0 \le t \le 1,$$

$$X_o = \psi(X_1),$$

where the diffusion coefficient σ_t is supposed to be deterministic, but the functions b_t and $\psi(x)$ may be random. More precisely, we will assume the following:

(H): (i) $\sigma \in L^2([0,1])$, and $|\sigma|_2 > 0$.

    (ii) $b : [0,1] \times \Omega \to R^1$ is a measurable process such that there is a constant C and a set $N_1 \in \mathcal{F}$ of probability one such that $|b_t(\omega)| \le C$, for all $0 \le t \le 1$ and all $\omega \in N_1$.

    (iii) $\psi : \Omega \times R \to R$ is a measurable function such that there exist a constant $K > 0$ and a set $N_2 \in \mathcal{F}$ of probability one verifying

$$\begin{aligned} |\psi(\omega, x) - \psi(\omega, y)| &\le K|x - y| \\ |\psi(\omega, x)| &\le K \end{aligned}$$

for all $x, y \in R^1$, and $\omega \in N_2$.

For stating the result of the existence and uniqueness of a solution of the stochastic differential equation (3.4.1) with boundary condition we introduce some notations. Consider the family of transformations $T_t, A_t : \Omega \to \Omega$, $0 \le t \le 1$, given by:

$$T_t\omega = \omega + \int_0^{t\wedge.} \sigma_s ds,$$

$$A_t\omega = \omega - \int_0^{t\wedge.} \sigma_s ds.$$

Note that $T_t A_t \omega = A_t T_t \omega = \omega$, $\omega \in \Omega$, and recall that the density L_t of T_t with respect to the Wiener measure is given by

$$L_t = \frac{dP \circ [T_t]^{-1}}{dP} = \exp\{\int_0^t \sigma_s dW_s - \frac{1}{2}\int_0^t \sigma_s^2 ds\}.$$

Set

$$Z_t = \exp\{\int_0^t b_s(T_s)ds\},$$

and, finally, define

$$(3.4.2) \qquad X_t = X_o(A_t)Z_t(A_t)L_t.$$

If the random variable X_o is known and bounded, the process (X_t) given by (3.4.2) verifies $(X_s) \in L^2([0,1] \times \Omega), (\sigma_s X_s I_{[0,t]}(s)) \in \text{Dom}\,\delta$ for all $0 \leq t \leq 1$, and

$$X_t = X_o + \int_0^t \sigma_s X_s dW_s + \int_0^t b_s X_s ds$$

(cf. Theorem 3.2.1).

Here we aim at extending this result to the case of the equation (3.4.1) with a random boundary condition of the form $X_o = \psi(X_1)$. Substituting X_1 given by (3.4.2) into this boundary condition equation provides

(3.4.3) $X_o = \psi(X_o(A_1)Z_1(A_1)L_1).$

First we will show that the equation (3.4.3) possesses a unique solution X_o.

Proposition 3.4.1 *Suppose that the functions σ, b and ψ satisfy the hypothesis (H). Then there exists a unique solution $X_o \in L^\infty(\Omega)$ of equation (3.4.3), and*

$$|X_o| \leq K, \qquad a.e.$$

Proof: Define $X_o^{(o)} = \psi(0)$ and

$$X_o^{(n+1)} = \psi(X_o^{(n)}(A_1)Z_1(A_1)L_1), \quad \text{for all } n \geq 0,$$

recursively. Using the Lipschitz property (H.(iii)) of ψ we obtain

$$|X_o^{(n+1)} - X_o^{(n)}| \leq Ke^C L_1 |X_o^{(n)}(A_1) - X_o^{(n-1)}(A_1)|, \text{ for all } n \geq 1,$$

and by iteration

$$
\begin{aligned}
|X_o^{(n+1)} - X_o^{(n)}| &\leq (Ke^C)^n L_1 L_1(A_1) L_1(A_1^2) \ldots L_1(A_1^{n-1}) |X_o^{(1)}(A_1^n) - X_o^{(o)}(A_1^n)| \\
&\leq 2K(Ke^C L_1)^n \exp\{-\frac{n(n-1)}{2}|\sigma|_2^2\}.
\end{aligned}
$$

This implies the existence of a random variable X_o such that

$$\lim_{n \to \infty} |X_o^{(n)} - X_o| = 0,$$

and, consequently,

$$\lim_{n \to \infty} |X_o^{(n)}(A_1) - X_o(A_1)| = 0.$$

Thus, passing to the limit in the iteration we obtain

$$X_o = \psi(X_o(A_1)Z_1(A_1)L_1),$$

and, in particular, we see that $|X_o| \leq K$, a.e. because of (H.(iii)). The uniqueness of a solution of (3.4.3) can be proved by means of a similar argument. Let X_o and Y_o be two solutions. Then, proceeding as above we get

$$
\begin{aligned}
|X_o - Y_o| &\leq (Ke^L L_1)^n \exp\{-\frac{n(n-1)}{2}|\sigma|_2^2\}|X_o(A_1^n) - Y_o(A_1^n)| \\
&\leq 2K(Ke^L L_1)^n \exp\{-\frac{n(n-1)}{2}|\sigma|_2^2\}, \text{ for all } n \geq 1.
\end{aligned}
$$

Since the right-hand side tends to zero as $n \to \infty$, we conclude that X_o and Y_o coincide.

The main result is the following theorem:

Theorem 3.4.2 *Suppose that the functions (σ_t), (b_t) and ψ satisfy the assumption (H). Then there exists a unique solution (X_t) of equation (3.4.1) inside the class of processes $(X_t) \in L^1([0,1] \times \Omega)$ such that $X_t \in L^1(\Omega)$ and $(\sigma_s X_s I_{[0,t]}(s)) \in Dom\,\delta$, $0 \le t \le 1$. This solution is given by (3.4.2) and (3.4.3) and is in $\bigcap_{p>1} L^p([0,1] \times \Omega)$.*

Proof (Existence): From Theorem 3.2.1 we know that the process $(X_t = X_o(A_t)Z_t(A_t)L_t)$ with X_o as the unique solution of equation (3.4.3) is a solution of the initial value problem

$$X_t = X_o + \int_0^t \sigma_s X_s dW_s + \int_0^t b_s X_s ds, \qquad 0 \le t \le 1.$$

Since, additionally, the boundary condition is satisfied, (X_t) is a solution of (3.4.1). Furthermore, taking into account

$$|X_o(A_t)Z_t(A_t)| \le Ke^C, \quad \text{a.e.}, \quad 0 \le t \le 1,$$

we see that (X_t) belongs to $L^p([0,1] \times \Omega)$ for all $p > 1$.
Proof (Uniqueness): Let $(Y_t) \in L^1([0,1] \times \Omega)$ be a solution of (3.4.1), i.e., $Y_t \in L^1(\Omega)$, $(\sigma_s Y_s I_{[0,t]}(s)) \in Dom\,\delta$, $0 \le t \le 1$ and (3.4.1) is satisfied. Clearly, since $Y_o = \psi(Y_1)$, the random variable Y_o is bounded by K, and so we can again apply Theorem 3.2.1 in order to deduce that $Y_t = Y_o(A_t)Z_t(A_t)L_t$, $0 \le t \le 1$. Putting $t = 1$ and substituting this expression for Y_t in the boundary equation we obtain

$$Y_o = \psi(Y_o(A_1)Z_1(A_1)L_1).$$

Now, according to Proposition 3.4.1 the solution Y_o of this equation is unique. This completes the proof.

Remarks:

1. The restriction of (H(i)) to $\sigma \in L^2([0,1])$, which requires that $|\sigma|^2 > 0$, is a natural one. If $|\sigma|_2^2$ vanished, we would get a purely determinstic equation with boundary equation, which is solvable in very special cases only.

2. The more interesting and technically much harder case of the nonlinearity of the drift part $b_s(\omega, X_s(\omega))$ in (3.4.1) endowed with a possibly unbounded boundary condition function ψ, as well as the Fréchet differentiability of the solution under the additional assumption that both b and ψ are deterministic $C^2(R^1)$-functions are studied in [16].

References

[1] Airault, H.; Malliavin, P.: Geometric Integration on the Wiener space. Bull. Sci. Math. II. Ser. 112 (1988), 3-52

[2] Bell, D.R.: The Malliavin calculus. Harlow: Longman 1987 (Pitman Monographs and Surveys in Pure and Applied Math., 34)

[3] Berger, M.A.; Mizel, V.J.: An extension of the stochastic integral. Ann. Probab. 10 (1982) 2, 435-450

[4] Buckdahn, R.: Quasilinear Partial Stochastic Differential Equations without Nonanticipation Requirement. Preprint 176, Sekt. Math., Humboldt-Univ. Berlin 1988

[5] Buckdahn, R.: Girsanov Transformation and Linear Stochastic Differential Equation without Nonanticipation Requirement. Preprint 180, Sekt. Math, Humboldt-Univ. Berlin 1989

[6] Buckdahn, R.: Anticipating linear stochastic differential equations. In: Zabczyk, J. (ed.) Stochastic Systems and Optimization. Proceedings of the 6th IFIP WG 7.1 Working Conference Warsaw 1988. - Berlin; Heidelberg; New York: Springer 1989, 18-23 (Lect. Notes Control Inf. Sci. 136)

[7] Buckdahn, R.: A linear stochastic differential equation with Skorohod integral. In: Langer, H.; Nollau, V. (eds.) Markov processes and control theory. - Berlin: Akademie-Verlag 1989, 9-15 (Math. Research, 54)

[8] Buckdahn, R.: Transformations on the Wiener space and Skorohod-type stochastic differential equations. Seminarbericht 105, Sekt. Math., Humboldt-Univ. Berlin 1989

[9] Buckdahn, R.: The nonlinear transformation of the Wiener measure. In: Dozzi, M.; Engelbert, H.J.; Nualart, D. (eds.) Stochastic processes and their topics. - Berlin: Akademie-Verlag 1991, 9-16 (Math. Research, 61)

[10] Buckdahn, R.: Anticipative Girsanov transformations. Probab. Theory Relat. Fields 89 (1991), 211-238

[11] Buckdahn, R.: Linear Skorohod stochastic differential equations. Probab. Theory Relat. Fields 90 (1991), 223-240

[12] Buckdahn, R.: Skorohod Stochastic Differential Equations of Diffusion Type. To apppear in Probab. Theory Relat. Fields

[13] Buckdahn, R.; Enchev, O.: Stochastic Calculus with Anticipating Integrands on the Abstract Wiener Space. Preprint 237, Sekt. Math., Humboldt-Univ, Berlin 1989

[14] Buckdahn, R.; Enchev, O.: Nonlinear transformations on the abstract Wiener space. Preprint 240, Sekt. Math, Humboldt-Univ. Berlin 1989

[15] Buckdahn, R.; Föllmer, H.: A Conditional Approach to the Anticipative Girsanov Transformation. Submitted to Prob. Th. Rel. Fields.

[16] Buckdahn, R.; Nualart, D.: Skorohod stochastic differential equations with boundary conditions. To appear in Stochastics and Stoch. Rep.

[17] Cameron, R.H.; Martin, W.T.: The transformation of Wiener integrals by nonlinear transformations. Trans. Am. Math. Soc. 66 (1949), 253-283

[18] Cruzeřro, A.B.: Equations différentielles ordinaires: non explosion et mesures quasi-invariantes. J. Funct. Anal. 54 (1983), 193-205

[19] Davydov, Yu. A.: On distributions of multiple Wiener-Itô integrals. Teor. Verojat. i Prim., 35 (1990) 1, 51-62

[20] Donati-Martin, C.: Equations différentielles stochastiques dans $I\!R$ avec contions aubord. Stochastics 35 (1991), 143-173

[21] Enchev, O.: Girsanov's Theorem for anticipative shifts. Preprint 1991

[22] Enchev, O.: Anticipative Diffusion and related change of measures. Preprint 1991

[23] Federer, H.: Geometric measure theory. Berlin; Heidelberg; New York: Springer 1969

[24] Gihman, I.I.; Skorohod, A.W.: Density of probability measures in functional spaces. Usp. Mat. Nauk 6 (1966), 83-156

[25] Gross, L.: Integration and nonlineaar transformations in Hilbert space. Trans. Am. Math. Soc. 159 (1971), 57-78

[26] Ikeda, N.; Watanabe, S.: Stochastic Differential Equations and Diffusion Processes. Amsterdam: North-Holland 1981

[27] Kunita, H.: First Order Stochastic Partial Differential Equations. - In: Taniguchi Symp. SA. Katata 1982, 249-269

[28] Kuo, H.H.: Integration Theory in Infinite Dimensional Manifolds. Trans. Amer. Math. Soc. 159 (1971), 57-78

[29] Kusuoka, S.: The non-linear transformation of Gaussian measure on Banach space and its absolute continuity (1). J. Fac. Sci., Univ. Tokyo, Sect. IA 29 (1982), 567-597

[30] Kusuoka, S.; Stroock, D.: Applications of the Malliavin calculus, Part II. J. Fac. Sci., Univ. Tokyo, Sect. IA, 32 (1985) 1, 1-76

[31] Meyer, P.A.: Probability and Potentials. Waltham: Blaisdell 1966

[32] Nualart, D.: Noncausal stochastic integrals and calculus. In: Korezlioglu, H.; Ustunel, A.S. (eds.) Stochastic Analysis and Related Fields. Proceedings, Silivri 1986. - Berlin; Heidelberg; New York: Springer 1988, 80 -129 (Lect. Notes Math., 1316)

[33] Nualart, D.; Pardoux, E.: Stochastic calculus with anticipating integrands. Probab. Theory Relat. Fields 78 (1988), 535-582

[34] Nualart, D.; Pardoux, E.: Boundary value problems for stochastic differential equations. Ann. Probab. 19 (1991) 3, 1118-1144

[35] Nualart, D.; Vives, J.: Continuité absolue de la loi du maximum d'un processus continu. C.R. Acad. Sci., Paris, Sér. I, 307 (1988) 7, 349-354

[36] Nualart, D.; Zakai, M.: Generalized stochastic integrals and the Malliavin calculus. Probab. Theory Relat. Fields 73 (1986), 255-280

[37] Ocone, D.; Pardoux, E.. Linear stochastic differential equations with boundary conditions, Probab. Theory Relat. Fields 82 (1989), 489-526

[38] Ocone, D.; Pardoux, E.: A generalized Itô-Ventzeli formula. Application to a class of anticipating stochastic differential equations. Ann. Inst. Henri Poincaré, Probab. Stat. 25 (1989), 39-71

[39] Pardoux, E.: Applications of anticipating stochastic calculus to stochastic differential equations. Preprint 1989

[40] Pardoux, E.; Protter, Ph.: Stochastic Volterra equations with anticipating coefficients. Ann. Probab. 18 (1990) 4, 1635-1655

[41] Ramer, R.: On non-linear transformations of Gaussian measures. J. Funct. Anal. 15 (1974), 166-187

[42] Shiota, Y.: A linear stochastic integral differential equation containing the extended Itô integral, Math. Rep., Toyama Univ. 9 (1986), 43-65

[43] Sussmann, H.: On the gap between deterministic and stochastic ordinary differential equations. Ann. Probab. 6 (1978), 19-41

[44] Ustunel, A. S.: Some comments on the filtering of diffusions and the Malliavin calculus. In: Korezlioglu, H.; Ustunel, A.S. (eds.) Proc. Silivri Conf. 1986. - Berlin; Heidelberg; New York: Springer 1988, 247-266 (Lect. Notes Math., 1316)

[45] Ustunel, A. S.; Zakai, M.. Transformation of Wiener measure under anticipative flows, Preprint 1990

Editorial Information

To be published in the *Memoirs*, a paper must be correct, new, nontrivial, and significant. Further, it must be well written and of interest to a substantial number of mathematicians. Piecemeal results, such as an inconclusive step toward an unproved major theorem or a minor variation on a known result, are in general not acceptable for publication. *Transactions* Editors shall solicit and encourage publication of worthy papers. Papers appearing in *Memoirs* are generally longer than those appearing in *Transactions* with which it shares an editorial committee.

As of June 7, 1994, the backlog for this journal was approximately 7 volumes. This estimate is the result of dividing the number of manuscripts for this journal in the Providence office that have not yet gone to the printer on the above date by the average number of monographs per volume over the previous twelve months, reduced by the number of issues published in four months (the time necessary for preparing an issue for the printer). (There are 6 volumes per year, each containing at least 4 numbers.)

A Copyright Transfer Agreement is required before a paper will be published in this journal. By submitting a paper to this journal, authors certify that the manuscript has not been submitted to nor is it under consideration for publication by another journal, conference proceedings, or similar publication.

Information for Authors and Editors

Memoirs are printed by photo-offset from camera copy fully prepared by the author. This means that the finished book will look exactly like the copy submitted.

The paper must contain a *descriptive title* and an *abstract* that summarizes the article in language suitable for workers in the general field (algebra, analysis, etc.). The *descriptive title* should be short, but informative; useless or vague phrases such as "some remarks about" or "concerning" should be avoided. The *abstract* should be at least one complete sentence, and at most 300 words. Included with the footnotes to the paper, there should be the 1991 *Mathematics Subject Classification* representing the primary and secondary subjects of the article. This may be followed by a list of *key words and phrases* describing the subject matter of the article and taken from it. A list of the numbers may be found in the annual index of *Mathematical Reviews*, published with the December issue starting in 1990, as well as from the electronic service e-MATH [telnet e-MATH.ams.org (or telnet 130.44.1.100). Login and password are e-math]. For journal abbreviations used in bibliographies, see the list of serials in the latest *Mathematical Reviews* annual index. When the manuscript is submitted, authors should supply the editor with electronic addresses if available. These will be printed after the postal address at the end of each article.

Electronically prepared manuscripts. The AMS encourages submission of electronically prepared manuscripts in $\mathcal{A}\mathcal{M}\mathcal{S}$-TeX or $\mathcal{A}\mathcal{M}\mathcal{S}$-LaTeX because properly prepared electronic manuscripts save the author proofreading time and move more quickly through the production process. To this end, the Society has prepared "preprint" style files, specifically the amsppt style of $\mathcal{A}\mathcal{M}\mathcal{S}$-TeX and the amsart style of $\mathcal{A}\mathcal{M}\mathcal{S}$-LaTeX, which will simplify the work of authors and of the production staff. Those authors who make use of these style files from the beginning of the writing process will further reduce their own effort. Electronically submitted manuscripts prepared in plain TeX or LaTeX do not mesh properly with the AMS production systems and cannot, therefore, realize the same kind of expedited processing. Users of plain TeX should have little difficulty learning $\mathcal{A}_{\mathcal{M}}\mathcal{S}$-TeX, and LaTeX users will find that $\mathcal{A}_{\mathcal{M}}\mathcal{S}$-LaTeX is the same as LaTeX with additional commands to simplify the typesetting of mathematics.

Guidelines for Preparing Electronic Manuscripts provides additional assistance and is available for use with either $\mathcal{A}_{\mathcal{M}}\mathcal{S}$-TeX or $\mathcal{A}_{\mathcal{M}}\mathcal{S}$-LaTeX. Authors with FTP access may obtain *Guidelines* from the Society's Internet node e-MATH.ams.org (130.44.1.100). For those without FTP access *Guidelines* can be obtained free of charge from the e-mail address guide-elec@ math.ams.org (Internet) or from the Customer Services Department, American Mathematical Society, P.O. Box 6248, Providence, RI 02940-6248. When requesting *Guidelines*, please specify which version you want.

At the time of submission, authors should indicate if the paper has been prepared using $\mathcal{A}_{\mathcal{M}}\mathcal{S}$-TeX or $\mathcal{A}_{\mathcal{M}}\mathcal{S}$-LaTeX. The *Manual for Authors of Mathematical Papers* should be consulted for symbols and style conventions. The *Manual* may be obtained free of charge from the e-mail address cust-serv@math.ams.org or from the Customer Services Department, American Mathematical Society, P.O. Box 6248, Providence, RI 02940-6248. The Providence office should be supplied with a manuscript that corresponds to the electronic file being submitted.

Electronic manuscripts should be sent to the Providence office immediately after the paper has been accepted for publication. They can be sent via e-mail to pub-submit@math.ams.org (Internet) or on diskettes to the Publications Department, American Mathematical Society, P.O. Box 6248, Providence, RI 02940-6248. When submitting electronic manuscripts please be sure to include a message indicating in which publication the paper has been accepted.

Two copies of the paper should be sent directly to the appropriate Editor and the author should keep one copy. The *Guide for Authors of Memoirs* gives detailed information on preparing papers for *Memoirs* and may be obtained free of charge from the Editorial Department, American Mathematical Society, P.O. Box 6248, Providence, RI 02940-6248. For papers not prepared electronically, model paper may also be obtained free of charge from the Editorial Department.

Any inquiries concerning a paper that has been accepted for publication should be sent directly to the Editorial Department, American Mathematical Society, P.O. Box 6248, Providence, RI 02940-6248.

Recent Titles in This Series

(Continued from the front of this publication)

(See the AMS catalog for earlier titles)